PREFACE

The following pages are the slightly expanded notes of a course of lectures which the writer has been giving to agricultural students for the past ten years. They constitute an introduction to plant and animal breeding, subjects of great and growing importance in modern agriculture.

The treatment is from a biological standpoint for the student who desires to specialize in animal husbandry, cereal husbandry or horticulture should have clear conceptions of the more important problems of organic evolution and of the main laws and theories relating to variation, heredity and related subjects. The discussion of the more direct application of the principles discussed has been left to the special departments.

The general public is taking an intelligent interest in many genetic problems, especially those relating to Eugenics. While the discussion may at times appear too technical for the average reader, it is believed that even these parts may be understood by him after a little careful study.

Frequent references are made to larger works which contain fuller and more extended discussions of many of the topics treated, and it is the writer's hope that this *Introduction*, condensed as it is, may be instrumental in persuading a larger number of persons to pursue their studies of this very interesting subject for its informational, educational and practical values.

The writer lays no claim to originality, except it may be in the presentation of the topics. He has had little opportunity for experimentation, but for many years his lot has been cast where breeding experiments have been carried on and important results secured.

In the new and rapidly expanding field of Genetics where results are being presented to the world by thousands of investigators it is extremely probable that many important phases have been overlooked, some misinterpreted, and others left obscure by reason of the condensed treatment.

The writer is under deep obligation to many friends who have had an opportunity of reading the manuscript for their criticism.

Macdonald College,
August, 1920.

W. LOCHHEAD.

An Introduction to
HEREDITY AND GENETICS
Modern Biological Laws and Theories
Relating to Animal and Plant Breeding

INTRODUCTION

Ever since the dawn of civilization attention has been given to the breeding of plants and animals, as evidence the numerous strains of cultivated plants and domesticated animals that now exist and have been in existence for many thousands of years. There seems to be no doubt that all of these domesticated forms have originated from wild forms. The origin of some of the forms is actually known, but that of the majority is lost in the centuries preceding the time of written records.

In most cases considerable differences now exist between the wild and the domesticated forms. The superiority of the latter over the former has been brought about by crossing and by intelligent selection of the most desirable forms, according to the purpose man had in mind.

Man's knowledge of the laws governing the production of animals and plants has been gradually acquired, slowly for a long time, but more rapidly as his acquaintance with the organic world increased. Especially have great strides been made during the last fifty years—since the appearance of Darwin's *Origin of Species* (1859), when, with overwhelming evidence, it was shown that organic beings are the "modified descendants of earlier forms, that in some way (by Natural Selection) new forms have arisen from the old ones, and have given rise, in turn, to other forms."

In the chapters that follow an attempt is made to outline briefly the more important theories that have been put forward by biologists to explain the development of plant and animal life, and to summarize the information gained by experimental investigation relating to inheritance. These will be discussed with the object of showing their bearing on the important problems of breeding. These chapters form, therefore, an introduction to the more technical courses given in the departments of Animal Husbandry including Poultry Husbandry, Cereal Husbandry or Agronomy, and Horticulture where direct applications are made to the better breeding of animals and plants.

Heredity has been defined as "the organic or genetic relation between successive generations," and again as "the germinal resemblance among organisms related by descent." Genetics is "the science which deals with the *coming into being* of organisms" by breeding carried on from generation to generation. Essentially Genetics has to do with the study of heredity, and its main object is to determine not only the mode of action of the germinal factors concerned in bringing about the relation implied in heredity, but also that of the external agencies that may affect the development of the new individual.

Genetics is breeding under rigid control so that we may know what is happening. At the present time the control is such that it is not possible to know all that is happening, but as Genetics is one of the youngest sciences, dating as it does from 1900 when Mendel's discoveries were made known, we may confidently look forward to a time when a method of continuous control may be employed in all breeding experiments.

We can probably all agree with Bateson when he says: "An exact determination of the laws of heredity will probably work more change in man's outlook on the world and in his power over nature than any other advance in natural knowledge that can be clearly foreseen."

Chapter 1—THE WORLD OF LIVING THINGS

It is very probable that ever since man's appearance on the earth the living things about him were more or less closely observed and their most evident likenesses and differences noted. In other words, our early ancestors made rough groupings of the animals and plants with which they came into contact. One of the first observers to record his observations was Aristotle (~~460–300~~ B.C.). His groupings were crude according to present-day standards as they were based mainly on external similarities of structure, but they remained practically unaltered for over 2,000 years, until the time of Ray (1628–1705) and Linnæus (1707–1778). Ray defined the term *species* but it was reserved for Linnæus to establish the binomial system of nomenclature and the grades of classification, viz., *Class, Order, Genus, Species* and *Variety* in his great work "Systema Naturæ."

The first edition of "Systema Naturæ" was published in 1735 and the last (12th) in 1768. Zoologists accept as the starting point for determining the generic and specific names of animals the tenth edition of 1758, while botanists have taken the 1753 edition of the "Species Plantarum", of which the first edition was published in 1737, as their starting point in nomenclature.

Before Linnæus, the lion was described as the "cat with a tuft at the end of the tail," and the tiger as the "yellow cat variegated with long black stripes." Such a method of description was both cumbersome and unsystematic, especially when the number of species became numerous.

Linnæus gave each species a double name, e.g., *Felis leo* and *Felis tigris*, the second name being the specific title, and the first the *genus*, a group of more or less similar species. In addition, genera were grouped into *Orders*, and Orders into *Classes*.

With the adoption of the binomial system "certainty and precision were introduced into the art of description."

Batsch in 1780 introduced the term *Family* between Genus and Order, and Haeckel in 1886 the term *Phylum* for a grouping of similar classes.

The successive steps are known as Divisions, Classes, Orders, Families, Genera, Species, Varieties which may be arranged as follows:

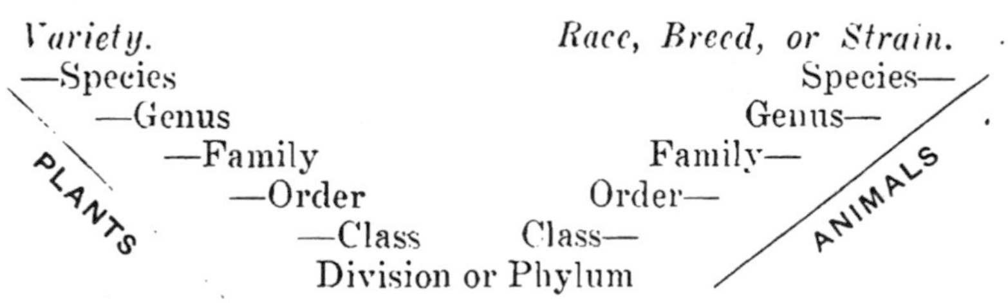

Thus in time through the efforts of many expert systematists all the known plants and animals have been *classified*. For example, the Spy apple belongs to the Division *Spermaphytes*, (seed-bearing plants), to the Class *Dicotyledons* (with two-seed leaves), to the Order *Rosales*, to the Family *Rosaceæ*, to the Genus *Pyrus*, to the Species *Mali*, and finally to the Variety *Spy*.

Similarly the domestic cat belongs to the Phylum *Chordata* (back-boned animals), to the Class *Mammalia* (mammals), to the Order *Carnivora* (flesh-eaters), to the Family *Felidæ* (the cat kind), to the Genus *Felis*, to the Species *domesticus*.

Ray defined a *species* as all similar individuals exhibiting constant characters from generation to generation, while, according to Linnæus, each species was descended from an originally created pair. "There are as many species", he said, "as issued in pairs from the Creator's hands," and each species expressed an idea in the Divine mind. This conception of species was generally held until the publication of the "Origin of Species" in 1859.

According to Darwin, however, classification meant the grouping of organisms to show relationships due to community of descent. He likened the affinities of all the beings of the same class to a great tree where the green and budding twigs represent existing species and those produced during former years the long succession of extinct species.

As a result, the Linnæan conception of species has been modified. Darwin and many investigators since his time have shown that Linnæan species are really groups of closely related forms, and that it is often impossible in nature to determine their limits on account of the large number of intermediate forms.

The idea of *species* is, therefore, an "abstract one: in nature there are no species—only individuals." The term, however, is still retained as a convenient one by systematists to designate all the individuals which agree in all essential features and live, as nearly as possible, under similar external conditions.

De Vries believes that his "elementary species" and "varieties" are the real units in nature. Individuals are considered as aggregates of "unit characters" which may combine in various ways, thus making it impossible to keep the group called a "species" as a definite entity.

The classification of organisms has, therefore, passed through four eras: 1. the *Pre-Linnæan*, 2. the *Pre-Darwinian*, 3. the *Pre-Mendelian*, and 4. the *Post-Mendelian.*

The Pre-Linnæan era (—1753) is characterized by crude unsystematic attempts at classification; the Pre-Darwinian era (1753–1859) by the introduction of system, frequently unnatural, and with distinct barriers between the members of each group; the Pre-Mendelian era (1859 1900) by the evolutionary conception of community of descent with the barriers between species more or less broken down; and the Post-Mendelian era (1900-) by the unit-character conception of individuals and the introduction of De Vries' 'elementary species' and 'varieties' as the real units in nature.

Chapter 2—THE ORIGIN AND EVOLUTION OF ORGANIC LIFE

(a)—Theory of Fixity of Species or Special Creation

The general opinion regarding the origin of plants and animals up to the time of Darwin was that every species came into existence by a distinct act of special creation. Each type or species remained distinct and was unable to vary in any of its individuals, except within very narrow limits. This theory was put into definite form by Suarez (1548–1617), a Spanish Jesuit, and was accepted as a doctrine of the Catholic Church. It was taken up by Milton in his "Paradise Lost," and accepted by Puritanism; it was accepted by Linnæus in his classification of plants and animals, and became "current intellectual coin."

It will be noted that this theory precludes all attempts to discover the origin of organic beings. Species were created, and nothing more was required to be said or done.

(b)—Theory of Evolution or Descent with Modification

The evolutionary conception of the organic world was held by Aristotle among the Greeks (~~469–399~~ B. C.), but for lack of facts his views were not clear and rather enigmatic. He expounded the doctrine of "a perfecting principle" which struggled with "the physical material cause," or matter itself, and worked out a continuous and progressive adaptation."[1]

But little addition was made during the following seven or eight centuries to the views held by Aristotle.

In the early centuries of the Christian era Augustine (353–430), influenced by Aristotle, held that creation was the institution of the order of nature.

During the Dark Ages science slept. After the Revival of Learning and the Reformation science again revived but men's thoughts were absorbed rather with the facts of nature than with the question of the origin of things.

Then followed: (1) a group of *Philosophic Evolutionists*—Bacon, Descartes, Leibnitz, Hume, Schelling, Kant and Hegel—who established the basis of the modern methods of studying the problem of evolution; and

(1)—"Nature produces those things which being continuously moved by a certain principle contained in themselves arrive at a certain end."

(2) a group of *Speculative Evolutionists*—Diderot, Bonnet, Robinet and Oken—who put forward a number of speculations regarding heredity and evolution that are often interesting on account of their crudeness.

Buffon (1707-1788) has been called the first of the great pioneers of modern evolution. His most noteworthy contribution was a theory of the direct action of the environment in the production of structural changes which are heritable.

Erasmus Darwin (1731–1802), grandfather of Charles Darwin, was another pioneer evolutionist who had fairly well defined ideas regarding the derivation of plants and animals. Unlike Buffon he believed that the environment only indirectly modified their organization. His *Zoonomia* was published in 1794.

The poet *Goethe* (1749–1832) was a firm believer in evolution, and contributed an important theory on the metamorphosis of foliar organs and another relating to an innate growth force. His views were published in 1794 and 1795.

Treviranus (1760–1837) was the forerunner of Charles Darwin in the emphasis he laid on the inter-relations between organisms and the environment. He was a Lamarckian with regard to the modifying influence of the environment.

Geoffroy St-Hilaire (1772–1844) was a pupil of Buffon, and shared his views concerning the direct action of the environment—published in 1828. He believed also that development might occur suddenly by leaps, and produce discontinuous variations.

Robert Chambers (1802–1871), the author of *Vestiges of the Natural History of Creation* (1844), prepared the minds of the reading public for Darwin's *Origin of Species*. His views of evolution, not always accurately expressed however, combined those of Buffon and St-Hilaire with that of Aristotle's idea of a perfecting principle (Read the Historical Sketch to Darwin's *Origin of Species* for a review of the progress of opinion on the origin of species).

The philosophic naturalists who did most to pave the way for the general acceptance of the doctrine of organic evolution were Lamarck and Darwin. They not only furnished strong evidences in support of the doctrine but also presented plausible theories regarding the mode of evolution.

Later investigators, notably Eimer, Weismann, Mendel and DeVries, have made valuable contributions, amending and enlarging upon the theories proposed by Lamarck and Darwin.

(c)—Lamarck[1] and Lamarckism

Lamarck (1744–1829), in his *Philosophie Zoologique*, published 1809, expounded more clearly than any of his predecessors a logical explanation of evolution. He recognized three factors:

(1) *a Changing Environment;*
(2) *the Effect of Use and Disuse; and*
(3) *the Inheritance of Acquired Characters;*

He reasoned as follows: Great changes in environment produce in animals new wants; new wants—changed habits; changed habits—the use of new parts or the disuse of old ones; and finally after many generations the production of new organs and the modification of old ones. He believed, however, that plants are modified directly by their environment. (Fig. 1)

Lamarck summarized his views in the following laws:

First Law.—"In every animal which has not finished its term of development, the frequent and sustained use of any organ strengthens and develops it, and increases it in size proportionate to the length of time it has been employed. On the other hand, the continued lack of use of any organ gradually weakens it until, at last, it disappears" (*Law of Use and Disuse*).

Second Law.—"Nature preserves everything she has caused the individual to acquire or lose through the influence of the environment to which its race has been for a long time exposed, and hence the predominance or loss of certain organs through the use or disuse. She does this by the production of new individuals which are endowed with the newly acquired organs, provided the acquired changes were common to the two sexes of the individuals that produced the new forms" (*Law of Inheritance*).

The essence of Lamarckism is, therefore, *adaptation* of plants and animals to their environment through *function*, or through the *use or disuse of organs*, and the *inheritance*

(1)—Lamarck had a varied career. He spent his boyhood at a Jesuit College in Amiens; served in the French army, studied medicine for four years; devoted himself first to Botany, later to Zoology. Other publications were: *Flora of France* (1778), *Natural History of Invertebrated Animals* (1815-1822). He reorganized the *Jardin des Plantes* (1793).

of functional changes. Lamarck's explanation of the snake's elongated legless body, and the duck's webbed feet is given by himself as follows:

"It is part of the plan of organization of reptiles, as well as of other vertebrates, that they have four legs attached to their skeleton. . . . but snakes acquired the habit of gliding over the ground and concealing themsleves in the grass; owing to their repeated efforts to elongate themselves, in order to pass through narrow spaces, their bodies have acquired a considerable length, not commensurate with their width. Under the circumstances, legs would serve no purpose, and consequently would not be used;

Fig. 1.—LAMARCK

long legs would interfere with the snakes' desire for gliding, and short ones could not move their bodies, for they can only have four of them. Continued lack of use of the legs in snakes caused them to disappear, although they were really included in the plan of organization of those animals."

On the other hand, "the frequent use of an organ, made constant by habit, increases the faculties of that organ, develops it and causes it to acquire a size and strength it does not possess in animals which exercise less. A bird, driven through want to water, to find the prey on which it feeds, will separate its toes whenever it strikes the water or wishes to displace itself on its surface. The skin uniting the base of the toes acquires, through the repeated separating of the toes, the habit of stretching; and in this way the broad membrane between the toes of ducks and geese has acquired the appearance we observe today."

Darwin fell back occasionally upon Lamarck's doctrine in explanation of adaptation and of the origin of new species, but included it only as a minor factor. The theory of the action of *use* and *disuse* explains better than any other (1) the origin of many indifferent characters such as change of color-patterns of butterflies' wings due to changes of temperature; (2) many simple adaptations of active organs, such as the development of muscles and bone crests; and (3) some simple adaptations of passive organs, such as the loss of hair and the layer of fat in the skin of whales.

On the other hand Lamarckism does not explain satisfactorily (1) many characters of active adaptation, such as the penetration of the lung-sacs of birds into the bones; (2) many complicated adaptations of active organs, such as eyes, smelling organs, auditory organs, light-making organs; and (3) all complicated passive adaptations such as mimicry (Plate).

According to Lamarck "the structure of organisms is in harmony with the conditions under which they live; in other words it is adapted to these conditions. The organism is shaped by the environment; usage develops the organs; without usage they atrophy. The modifications thus acquired are transmitted to posterity."

Lamarckism as a scientific theory of evolution is in one sense more complete than Darwinism in that it looks to the very cause of the change of organisms by its method of explaining adaptation.

(d)—Darwin and Darwinism

Charles Darwin (1809–1882)[1] published his "*Origin of Species by Natural Selection*" in 1859 after twenty years of most patient and painstaking labor. This work at once compelled the attention of scientists by its wonderful com-

(1)—Darwin was the grandson of Dr. Erasmus Darwin. He studied at Edinburgh and Cambridge, then served as Naturalist on the exploring ship **Beagle** (1831-1836). He settled at Downs in England on his return and began that wonderful series of researches on plants and animals which he has described in a large number of well written volumes.

pilation of facts bearing on the question of evolution, and by the masterly arguments for the origin of species through the agency of Natural Selection. According to Darwin and Wallace (co-discoverers of the Theory of Natural Selection) there are three main factors in the evolution of species:

(1) *Over-production among organic beings;*

(2) *Variation; and*

(3) *Heredity.*

The main argument may be summarized as follows: As organic beings tend to increase in a geometrical ratio, while the means of subsistence tend to increase only in an arithmetical ratio (Malthus),[1] there naturally arises a severe struggle for life at some age, season or year. As there is much variation in the structure of even the most closely

Fig. 2.—CHARLES DARWIN

allied forms those forms that possess useful variations will most likely survive in the struggle for existence[2] in an environment of infinite complexity, both with regard to the relations of beings to one another and to the conditions of life: then because of the "strong principle of inheritance

(1)—Malthus' essay on "Population" was published in 1798 and the conclusions were based on data relating to human population obtained from many countries.

(2)—Darwin said: "I use this term ('struggle for existence') in a large and metaphorical sense, including the dependence of one being on another, and including (which is more important) not only the life of the individual, but success in leaving progeny."

the survivors will tend to produce offspring similarly characterized." Thus after many generations forms will arise which differ so much from the original forms as to be designated new species. The principle of preservation or survival of the fittest Darwin called *Natural Selection*. Natural Selection "leads to the improvement of each creature in relation to its organic and inorganic conditions of life, and consequently, in most cases, to what must be regarded as an advance in organization." (Fig. 2).

Darwin devoted a considerable portion of the "*Origin of Species*" to what he called "Evidences of Organic Evolution." These were considered from the historical standpoint in Chapters X to XIV under the following titles: "On the Imperfection of the Geological Record," "On the Geological Succession of Organic Beings," "Geographical Distribution," "Mutual Affinities of Organic Beings;" "Morphology: Embryology: Rudimentary Organs." Regarding the Imperfection of the Geological Record Darwin claimed that on account of the many gaps in the record of life preserved in the rocks, it would be hardly possible ever to expect a full knowledge of the organic life that had existed on the earth. Geological evidence, however, as it has accumulated since Darwin's day, is in strong accord with the theory of Evolution. Take for example the history of the ancestors of the horse in North America, as worked out by Marsh and Osborn, and the Archæopteryx, the earliest known bird with reptilian characters, etc., etc. Moreover, the geological record shows the gradual appearance of higher and higher forms. The Invertebrates appeared first, then fishes, amphibians, reptiles, mammals and birds in succession.

The present distribution of animals and plants over the surface of the earth can be most satisfactorily explained by the theory of evolution. The facts suggest the gradual dispersal of races from a starting point. Many factors, however, enter into the problem, such as climate and climatic changes; oceanic, desert, and mountain barriers; isolation, etc.

Darwin found that each island of the Galapagos group had its characteristic animal life, but the species on one island are closely similar to those on another, and to those on the adjoining continent. Moreover, the life on the larger central islands is more closely related than is the life on the more isolated islands. All these facts suggested to Darwin that the corresponding species on the island and the continent are related by a common descent.

On the pampas of Argentina Darwin noted the structural correspondence between such living forms as the sloths and ant-eaters, and the extinct Megatherium and Glyptodon whose fossil remains he had collected. To him the explanation was blood relationship.

A study of organic life, therefore, shows relationships, some near, others more remote. It can be compared best to a great tree which branched near the base into two large trunks—the animals and the plants—each dividing again into smaller and smaller branches showing closer and closer relationship at the tips. From this we get our conception of *classes, orders, families,* and *genera.*

The evidence of evolution from comparative structures is strong. The homology of organs in the vertebrates finds no interpretation except on the basis of relationship. The different sets of organs are apparently constructed on the same type. In the course of time the organs have been modified to become better fitted for different kinds of work; for example, the arm of a frog, the paddle of a turtle, the wing of a bird, the fore-leg of a horse, the wing of a bat, the flipper of a whale, and the arm of a man are homologous.

Sometimes organs that were probably at one time useful and performed functions are now functionless, *e.g.*, vestigial or rudimentary organs, such as the appendix in man, teeth in embryo of the whale-bone whale, ear-moving muscles, etc.

Darwin likened "vestigial structures" to the unsounded letters in many words, such as "o" in leopard, "b" in doubt, and "g" in reign, which though functionless tell us something of the history of these words.

Embryological evidence is also of much weight. The individual during its development (*ontogeny*) shows many traces of its racial development (*phylogeny*). Prof. Marshall said: "The individual climbs up its own genealogical tree." or, more technically expressed, "ontogeny tends to recapitulate phylogeny" (The Biogenetic Law of Haeckel). The young stages of the higher vertebrates, for example, show remarkable resemblance, and gill-slits occur in the embryos of reptiles, birds, and mammals.

The recapitulation theory as first enunciated was based on the false view that "in the course of evolution new end stages are added to the ontogeny as it had previously existed." Bu it was shown that "evolutionary changes may affect any portion of the life history, and therefore the course of ontogeny is not a sure indication of the course of phylogeny". Haeckel called those ontogenetic stages which repeated the phylogeny, "palingenetic," and those which did not, "coenogenetic." Conklin says: "ontogeny recapitulates phylogeny not in all its stages and forms, but in its *factors* and *principles*."

Isolation. By some evolution-philosophers *isolation* is given a prominent place as a factor in the evolution of new forms, but it is clear that it is not so important as selection and variation, Isolation tends to divergence from the parent forms, and, acting for long periods, develops prepotency or stability of type.

By isolation, it is clear, free intercrossing of members of a species is prevented. Romanes believed that organic evolution is impossible without isolation, but he undoubtedly laid too much emphasis on the swamping effect of free intercrossing on germinal variations.

"Isolation tends to the segregation of species into subspecies, makes it easier for new variations to establish themselves, promotes prepotency and fixes characters."

Many examples of the truth of the above statement might be cited from every division of the animal world, but it will suffice to instance the part of isolation in the development of races, such as the Nordic, Alpine and Mediterranean peoples of the world who maintain their characteristics of physical type and of character. It is likely, further, that the characteristic Scot, Irishman, and Welshman are largely the result of the action of isolation.

The breeds of cattle have arisen largely through artificial isolation, when close inbreeding was practised.
(Consult writings of D. S. Jordan, Romanes, Wagner and Gulick; also Madison Grant's work "The Passing of the Great Race.")

In any study of Isolation the following types of barriers must be considered:

(1) *Geographical barriers*, such as mountain ranges, oceans or seas, deserts, etc.

(2) *Temporal barriers*, where members of a species reach sexual maturity at different times of the year

(3) *Habitudinal barriers*, where a species splits into two or more castes with different habits of life.

(4) *Physiological barriers*, such as variations in form of the reproductive organs, preventing mating.

(5) *Psychological barriers*, due to profound antipathies.

Artificial Selection.—Darwin was much impressed with the results of artificial selection or selective breeding that man has effected among domesticated animals and cultivated plants. There is evidence that the nearly 200 well-marked breeds of domestic pigeons have all descended from the blue rock-pigeon (*Columbia livia*)[1], and the various breeds of poultry from the Indian jungle fowl (*Gallus bankivus*). The various breeds of horses, rabbits and ducks too have each probably come from a single wild species. On the other hand the breeds of dogs and cattle have arisen from more than one species.[2] (See page 33).

The origin of most of our garden flowers, fruits and vegetables is lost in obscurity, but many new forms are being evolved before our very eyes.

Darwin reasoned that if man has been the agent in evolving all these forms in a relatively short time, Nature is capable of effecting much also in a very long time.

Kinds of Evolution.—Evolution has taken place in many directions. There has been *convergence* as well as *divergence* in development, just as there has been *regressive* development as well as *progressive* development. Divergent progressive evolution is the type usually considered and represented in our genealogical trees. Examples of convergent development are the swift and the swallow, which are birds very distantly related; the bird and the bat, etc. (Read Willey's *Convergence in Evolution*).

Other cases where analogies are developed are *Typhlops*, a burrowing snake, *Amphisbæna*, a burrowing lizard, *Siphonops*, a burrowing amphibian, which are more or less worm-like; *Pteromys*, a flying squirrel, and *Petaurus*, a flying phalanger; the porcupine (*Erethizon*), a rodent, the hedge-hog (*Erinaceus*), an insectivore, and the spiny ant-eater (*Echidna*) a monotreme, all spine covered animals; also the eyes of Molluscs, vertebrates and Polychaete worms.

Correlation in Evolution.—"A variation important in the present may bring in its train one that is designed to be important in the future, and a variation too small in itself to be of value may be carried over the dead point into effect-

(1)—Some of the common breeds of domestic pigeons are the tumbler, pouter, fan-tail, carrier, homing, barb, trumpeter, common mongrel, archangel, fairy swallow, blackwinged swallow and bluetts.

(2)—The student of organic evolution should read and re-read Darwin's Origin of Species and Variation of Animals and Plants under Domestication. The latter is not as well known as the former, but in most respects it is equally important.

iveness because it is correlated with another variation of greater vital value" (Geddes and Thomson, *Evolution*).

Organic Selection.—"If the variation favored by the environment coincides with an *innate* variation of similar nature, the effects of both variations are more likely to make themselves felt than either separately. It may happen that a slight innate variation, too insignificant to serve a useful purpose is somehow amplified by an acquired variation of similar nature which adds itself to it and the two variations in combination form a proper basis for natural selection."

Sexual Selection.—To account for the occurrence of marked secondary sex-characters in many males Darwin proposed the supplementary theory of *Sexual Selection.* "This form of selection depends on a struggle between the individuals of one sex, generally the males, for the possession of the other sex. The result is not death to the unsuccessful competitor, but few or no offspring." There are two modes of this sexual selection, the fights between rival males and the preference the female shows for "the male whose *tout ensemble* has most successfully excited her sexual interest." (See also page 52 for Darwin's theory of Pangenesis).

Criticisms of the Natural Selection Theory.

Criticisms of this Theory have been very severe, especially by theologians about 40 years ago. The following are among the most important from a scientific standpoint:

1. The two types of variations are not clearly differentiated, and it is believed that fluctuating variations have a limit beyond which they cannot pass by continuous selection.
2. It is also believed that *acquired characters* are not heritable.
3. It is hard to understand how Natural Selection favors organisms with variations too small to be of any advantage. The theory does not explain the origin of useful variations. "It may explain the survival of the fittest but it cannot explain the arrival of the fittest."
4. It does not explain sterility between most species.
5. "Many characters useless for the preservation of individuals have been developed by one sex of a given species," *e.g.* the plumage of many male birds.

6. It demands almost unlimited time for the evolution of the higher forms from the simplest on the earth, which demand physicists and geologists are unwilling to grant.
7. Jordan, the French botanist, discovered more than 200 distinct forms or elementary species of *Draba verna*, each preserving its own special characters for generations. He frequently found a certain habitat of very limited extent occupied by several of these species forming an association. It is highly improbable that each of these species arose as an adaptation to the same environment. It would seem then that "the specific characters of *Draba verna* cannot have been evolved in the struggle for existence, that they are not adaptive characters, and that they are in themselves useless" (Jost).

It will be noted that Darwin's theory fails to answer satisfactorily the meaning of the different kinds of variattions, their rate of inheritance, or the mechanism of inheritance.

Fact and Mode of Organic Evolution.—One must be careful to distinguish between the general *fact* of organic evolution and the *mode* of evolution, or the method by which evolution has taken place. "Stated concretely, the general doctrine of evolution suggests that the plants and animals now around us are the results of natural processes working throughout the ages, that the forms we see are the lineal descendants of ancestors on the whole somewhat simpler, that these are descendants from yet simpler forms, and so on backward till we lose our clue in the unknown vital events of Pre-Cambrian ages, or in other words, in the thick mist of life's beginnings. The general idea of evolution is, therefore, that the present is the child of the past and the parent of the future" (Thomson, *The Study of Animal Life*).

The theory of organic evolution seeks to explain the world of genera, species and varieties, the progressive organization from the simple to the most complete, and the adaptations of living beings.

(e)—**Orthogenesis**

The origin of new forms as the result of a persistent determinate variation is called Orthogenesis. While organisms as a rule tend to vary in every direction, variation being indeterminate, yet there are many examples of organisms that have evolved along definite lines, and where the variation has apparently been predetermined. In some instances these variations developed to such an extent that they became disadvantageous, for example, the unwieldy Cretaceous reptiles, the coils of the Ammonites, the tusks of the mammoth, the giant antlers of the Irish stag, and the sabre-toothed tigers, among extinct animals.

Coulter says: "The history of such a group as Gymnosperms shows a tendency to vary in certain definite directions that have persisted from the early Paleozoic to the present time." We also see Orthogenesis probably in the odd pinnateness of the leaves of the sumacs, and in the abrupt pinnateness of the leaves of the partridge-pea, etc. There is a "racial tendency toward some particular line of development." Overspecialization can be explained by this theory—for example, the tusks of the wild boar, the giant horns of many wild sheep and goats, the enormous beaks of several rhynchophorous beetles, etc.

The explanation of this persistent variation still eludes us. Nageli (1884) explained Orthogenesis by means of "an inner directive force," an inherent force in the organic world that makes for progressive development.[1] Eimer (1888) on the other hand believed that Orthogenesis is produced and controlled by external factors of climate, food supply, and environment generally. Many biologists believe that Natural Selection, Mutation, and Orthogenesis may all be operative in producing new forms (Kellogg, *Darwinism To-day*).

Discussion:—Does Natural Selection play any part in Orthogenesis?

(f)—Protoplasm and the Origin of Life

Huxley very properly and aptly termed protoplasm "the physical basis of life" for everywhere life is associated with protoplasm. Evolutionists, therefore, in seeking for the origin of life have as their fundamental problem the origin of protoplasm. According to the mechanistic school, the problem is one belonging wholly to physics and chemistry, but the vitalistic school sees the working of a directive force, termed *vitalism*, in connection with the physico-chemical changes.

In the present state of science the solution of the problem should not be expected. Our knowledge of protoplasm is very incomplete on account of its organic com-

(1)—Contrast this theory with that of Aristotle's "perfecting principle."

plexity; moreover, the intimate structure of the protein compounds composing it is still unknown, not to speak of the physical conditions essential for the union of these compounds in the formation of the "*living stuff*."

Recent progress in the study of colloids suggests a time very remote in the earth's history when conditions of temperature, pressure and moisture were very different from those of the present, and perhaps more favorable for the production of colloids, the basis of protoplasm. From simple materials more and more complex substances would be gradually formed, if the analogy of the structure of the inorganic world in azoic times is any guide to an understanding of the changes that occurred. It is even believed by some that the making of protoplasm may be in operation under certain conditions at the present day, but we are not in a position to recognize it as such.

The vitalists maintain that there is a great gulf between mere colloidal substances and living protoplasm. The latter is characteristically unstable, and is "self-repairing, self-preservative, self-adjusting, self-increasing, self-stoking and self-reproducing"—characters which do not belong to any non-living machine.

The solution of the problem of the origin of life, if at all possible, belongs to the future. (Consult Osborn's "The Origin and Evolution of Life.")

The most that can be said at the present time is that life comes from life.

Chapter 3—VARIATIONS: KINDS, CAUSES

Before discussing the theories of Weismann and DeVries we shall consider *Variations*. That living things are variable is evident to most persons. Close observation shows that aside from the larger differences that exist between animals and plants belonging to different genera, families or orders, animals and plants produced by the same parents are not exactly alike. Darwin recognized two kinds of heritable variations:

(1) Small "individual variations" that normally occur among members of the same species [1] and

(2) Sports or "single variations" that sometimes occur among both plants and animals.

He believed also that those variations produced by the environment and functional changes may sometimes be

(1)—Darwin evidently recognized individual variations that are heritable and varieties that are the result of external influences—somatic modifications and not heritable.

transmitted to the offspring. His general conclusion was that heritable variations are due to external conditions acting directly on the germ-plasm, and that one of the most important external conditions has been the food supply (*Variation af Plants and Animals*).

Variations have very appropriately been termed" the raw materials of evolution," for if there were no variations there could be no evolution. The production of the highly-developed domesticated breeds of animals and plants of to-day is the result of man's conscious selection of those forms that showed variations in directions that suited his fancy.

Darwin supposed that organisms under domestication tend to vary far more than similar organisms do in the wild state, a result due, he thought, to more varied surroundings. Later investigators like Bateson inform us, however, that variation is as common among wild animals as it is among domesticated.

During the last forty years important progress has been made in the study of variations. Quetelet, a Belgian investigator, showed that ordinary variability follows the law of probability, and this conclusion has been frequently verified by later investigators. Whenever a large number of organisms of a kind are compared, and similar parts are carefully measured or counted, it is seen that the variations in any single character are very numerous. The larger variations from the average type in either direction are fewer than the smaller variations, and if these variations are plotted in paper, according to their size and frequency, a *curve of frequency* is obtained which corresponds to the law of probabilities. This curve has a crest decreasing rapidly to zero on either side.

According to this curve variability is limited and tends to return to the average condition. (See page 25).

Variations that behave in this manner are termed *fluctuating variations.* since they fluctuate about their average or mean.

The other kind of variations, viz., *discontinuous variations*,[1] has been shown by Bateson[2] (*Materials for*

(1)—Galton many years before illustrated the difference between continuous and discontinuous variations by the polygon.

(2)—Prof. W. Bateson, the noted English biologist and geneticist, was born in 1861 and educated at Rugby and Cambridge. He has filled the professorship of Biology at Cambridge, and of Physiology of the Royal Institution, and is now Director of the John Innes Horticultural Institution, Merton Park, Surrey. He was awarded the Darwin Medal in 1904, and was President of the B.A.A.S., 1914. His publications are—"Materials for the Study of Variation' (1894), "Mendels Principles of Heredity" (1902) and "Problems of Genetics."

the Study of Variations, 1894) to be quite frequent in occurrence, and by DeVries and others to be of great importance in the development of new species. Such variations are termed *mutations* by DeVries, and will be discussed at greater length under the head of *Mutation Theory*.

Kinds of Variations.—In the study of these two classes of variations four different kinds are now recognized:

1. *Morphological*, relating to differences in form or size:
   (*a*) Variations in size, of apples, horses, sheep, etc.
   (*b*) Variations in relative proportions of parts, *e.g.*, wider face, longer legs, etc.
2. *Substantive*, relating to differences in quality of the structure:
   (*a*) Variations in fineness of bone, firmness of muscles, etc., of animals.
   (*b*) Variations in quality of milk, brain, etc.
   (*c*) Variations in quality of apples, sugar beets, corn, etc.
   (*d*) Variations in color, immunity to disease, hardiness, etc.
3. *Meristic*, relating to deviations in pattern, expecially to repeated parts:
   (*a*) Variations in symmetry, either radial or bilateral, in double flowers, in four-leafed clovers.
   (*b*) Extra teats in cows, extra fingers or toes in man, etc., etc.
4. *Functional*, relating to deviations in the normal activity of the various organs and parts of the body or the plant:
   (*a*) Variation in the degree of activity of normal functions between different individuals of same species, *e.g.*, in milk secretion, meat production, speed, resistance to disease, and vitality generally.
   (*b*) Variations within the same individual, *e.g.* daily fluctuations, influence of age, use and disuse, and food upon functional activity.
   (*c*) Modification of normal functions by external or other influences, *e.g.*, galls, etc.
   (*d*) Normal functions exercised under abnormal conditions.

Curves of Variation.—The graphic representation of variations is often both instructive and suggestive. Use is made of "graph" paper (paper ruled into fine squares) in the plotting of "frequency curves." The scale used in plotting will depend on the number of cases to be covered.

The following curve represents the variations in the length of 448 seeds of bean (DeVries):—

Length in mm.	8	9	10	11	12	13	14	15	16	Total
Frequency.	1	2	23	108	167	106	33	7	1	448

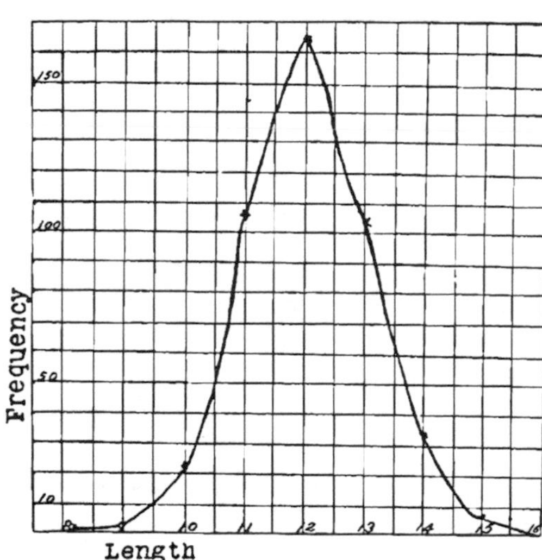

Fig. 3.—Frequency of Curve of the Variations in Length of 448 Seeds of Beans.

It will be noted that the curve is a normal curve symmetrically developed on both sides. Such a curve of variability expresses the fact that a large proportion of the beans are near the average length of 11 to 13 mm., those differing from this being relatively few on either side of the average. The *mode* of the curve is 12 or the group containing the largest number of individuals of a kind.

The following table represents the variability of milk production of 1,200 cows (Davenport):

Pounds milk.	1,500	2,500	3,500	4,500	5,500	6,500	7,500
No. of cows.	10	59	178	256	253	209	131

Pounds milk.	8,500	9,500	10,500	11,500	12,500	Total
No. of cows.	63	28	10	2	1	1,200

The curve would show clearly that the majority of the cows yield between 4,500 and 6,500 lbs. of milk per year.

The *average* length of the beans is ascertained to be 12 mm., and the *average* milk production 5.515 lbs.

Deviation.—The variability, or deviation from the average, gives us information as to the range of the deviation. There are two methods of measuring this deviation, the *Average* Deviation and the *Standard* Deviation. The average deviation of the beans is calculated as follows (12 being average length):

$1\ (12-8) + 2\ (12-9) + 23\ (12-10) + 108\ (12-11) + 167\ (12-12) + 106\ (13-12) + 33\ (14-12) + 7\ (15-12) + 1\ (16-12) \div 448$; which gives .8mm.

Formula: Average deviation = $S\frac{d.f}{n}$

Similarly, the *average deviation* of the milk production in the 1,200 cows can be shown to be 1,787.3 lbs. (5,515 lbs. being average production).

The *standard deviation*, considered by mathematicians to be more convenient as a measure of variability, is calculated in a slightly different way. In the beans it would be worked out as follows:—
$1\,(12-8)^2 + 2\,(12-9)^2 + 23\,1(2-10)^2 + 108\,(12-11)^2 + 167\,(12-12)^2 + 106\,(13-12)^2\,33\,(14-12)^2 + 7\,(15-12)^2 + 1\,(16-12)^2 \div 448$; and the square root of this quotient gives 1.1 mm.

Formula:—*Standard Deviation* = $\sqrt{\frac{S\,d^2.f}{n}}$

In the case of the production of milk from 1,200 cows the standard deviation is 1,770.1 lbs.

For the sake of clearness, the results are summarized in the following table.

A Comparison of Average and Standard Deviations as to Length of Beans

V	*f*	V*f*	V–M	*f* (V-M)	(V–M)²	*f* (V-M)²
8	1	–8	–4	–4	16	16
9	2	18	–3	–6	9	18
10	23	230	–2	–46	4	92
11	108	1188	–1	–108	1	108
12	167	2004	0	0	0	0
13	106	1378	1	106	1	106
14	33	462	2	66	4	132
15	7	105	3	21	9	63
16	1	16	4	4	16	16
	488	5409		361		551

V= Value, f= frequency, M= Mean or average.

M or Average = $\frac{5409}{448} = 12;$

Av. Dev. = $\frac{361}{448} = .8;$

St. Dev = $\frac{551}{448} = 1.1$

PROBABLE ERROR.—In statistical studies of variations the results are usually checked up by a determination of the **Probable Error.** (See E. Davenport and Babcock and Clausen for details.).

Co-efficient of Variation.—It is often important to compare the variability of different characters, by expressing the *Standard Deviation* as a percentage of the average. In the case of the beans the *co-efficient of variation* as to length is

$$\frac{1.1}{12} \times 100 = 9.16.$$

In the example of milk production cited above it is

$$\frac{1,770.1}{5515} \times 100 = 32.1$$

When the type of curve, the average and the standard deviations and the co-efficient of variation of a character are known, the nature of the variation may be said to be fairly well described.

Bimodal Curves.—Sometimes curves show two modes instead of one; such curves indicate either that the group is breaking into two types or that two distinct types have become mixed.

Skew Curves—Sometimes curves are unsymmetrical, and the mode lies considerably to one side of the average. Such curves show the direction in which the variations are tending.

Exercises.—Find examples of organisms that show (1) bimodal, (2) skew curves.

Additional Exercises in Variations

1. Plot the Curve and determine the Average and Standard Deviations, and the Co-efficient of Variation in the following analysis of sugar beets, after DeVries:—

Per cent sugar ..	12	12½	13	13½	14	14½	15	15½
Frequency......	340	635	1192	2205	3579	7178	5561	7829

Per cent. sugar........	16	16½	17	17½	18	18½	19
Frequency............	6925	4458	2233	692	133	14	5

2. Plot the curve and determine the Deviations and Co-efficient of Variation in the following study of length of nuts, after Watson:—

Length.........	30	34	38	42	46	50	54	58
Frequency......	2	7	28	59	49	33	6	1

3. Plot the curves and determine the Deviations in the number of ribs in the leaves of two beech trees, after Walter:

No. of Ribs.....	13	14	15	16	17	18	19	20
Frequency, 1st..	–	–	1	4	7	9	41	1
2nd tree........	3	4	9	8	2	–	–	–

4. Plot the curve and determine the Deviations for the number of ray flowers of bur marigold, after Needham:—

No......................	3	4	5	6	7	8	9	10	11
Frequency................	2	3	8	19	52	221	9	0	1

Bud or Clonal Variations

Sometimes buds arise on plants that produce branches differing markedly from the other branches. Such are called bud variations, and are often found on roses, coleus, carnations, nephrolepis, and chrysanthemums.[1] Some bud variations are undoubtedly mere fluctuations, but many are distinct mutations. Both types can be used in breeding—*e.g.*, violets, potato, and rose. Webber has shown that the various types of oranges, such as *nobilis, aurantium, decumana, limonum, myrtifolia*, and *trifoliata* have arisen as bud mutations from a single original form. A study of these types show Progressive, Retrogressive, and Regressive forms. Not much is really known about the causes producing bud variations, but probably their origin is due to the same causes that produce the mutations occurring in sexual reproduction.

The *Washington Navel* or *seedless orange* arose as a bud sport found at Bahia, Brazil, in 1826, and brought to Washington about 45 years ago.

Causes of Variations

Darwin said (*Origin of Species*): "Our ignorance of the laws of variation is profound. Not in one case out of a hundred can we pretend to assign any reason why this or that part has varied." This statement still holds true today. Some biologists maintain that variability is a primary property of organisms while others regard variations as the result of causes either extrinsic (Lamarckian) or intrinsic to the organism. For convenience, the causes of variation may be classified as (1) Internal, and (2) External.

(1)—Read Stout on Bud Selection in Coleus (Publ. 218 Carnegie Inst.) Inst.) and Coit's Citrus Fruits.

The **Internal Causes** are:

(1) *Cell Division.*—As growth is the result of cell division, and as all cell divisions involve very intricate divisions and re-arrangements of the chromosomes of the nucleus (see chapter 16) and the smaller determinants or physiological units, it would be strange if differences did not arise in the process due to both internal and external factors. We know that abnormalities in growth do occasionally rise from abnormal shuffling of the units concerned. We are not yet acquainted with the factors that determine the limits of growth, but it is proable that under certain conditions the limit of growth is reached sooner in some cases than in others. This in itself would produce variation.

(2) *Bisexual Reproduction.*—In the higher plants and animals a new organism is the result of the mingling of two inheritances in the male and female germ cells (*amphimixis*), and as a result must lead to the production of variation. The offspring, therefore, *tends* to likeness with the parents.

(3) There is much scope for the variation in the intricate processes represented in maturation and reduction. The loss of chromatic material allows for different combinations of the hereditary qualities. (See chapter 16)

(4) Recent studies of the Slipper-animalcule (*Paramecium aurelia*) by Woodruff and Erdmann reveal periodic nuclear changes of a breaking-down and a re-organizing nature. This process is called *endomixis*, and is believed to be regulatory in character as it secures the life of the race indefinitely in the absence of conjugation. If such a process occur in the many-celled animals may not the changes induce variations even in pure lines?

Again in the regeneration of a Planarian from a fragment, Child finds that a *rejuvenescence* takes place, as measured by the increased metabolism and power of resistance. May not rejuvenescence take place in the germ cells with every new generation at the time of re-constitution? May not such changes lead to variations?

(5 Vernon showed that the degree of ripeness of ova and sperms of sea-urchins influences the size of the larvae. Moreover, Weismann believed that a struggle goes on in the germs (germinal selection) between the *determinants*, the stronger and better nourished determinants determining the character of the germs, and the zygote (see page 55).

(6) Webber has shown that when white and yellow corn are allowed to cross the white ears will have many yellow grains. This phenomenon is due to the fertilization of the endosperm-nucleus by the second sperm cell (the first one fertilizing the egg cell), with the result that the endosperm is influenced the first year by the yellow germ cell. This phenomenon is called Xenia. (See page 124).

(7) *Hormones or Chalones or* Internal Secretions.—Certain glands the thyroid, pituitary, thymus, suprarenals, pineal, ovaries and testes, and the pancreas secrete substances that have an important influence on the character of the organs. Some physiological chemists seem to see in Internal Secretions strong evidences against the inheritance of structural units in the germ.

The **External Causes** are:

(1) *General effect of locality.*—It is a matter of common observation that locality has an important influence on animals and plants with regard to their size, yield, quality, flavor, etc. The factors operating are numerous and complex, partly climatic, partly those relating to soil conditions, and partly those relating to food supply. In general it may be said that where conditions of life differ there will be corresponding variations.

(2) *Food.*—The influence of quality and quantity of food has long been recognized by breeders of plants and animals. The influence is seen directly during the young or growing period. When plants or animals are stunted by reason of unsuitable food or an insufficient amount of suit-

able food, they do not recover from the effects. The effect of food may be observed also in the bee-hive in the production of queens. The term "high living" expresses a condition which produces variations, often called *disorders*, detrimental to the best development of organisms whether they be plants or animals.

(3) *Moisture.*—This factor in variation is obvious to every observer. Botanists classify plants according to the amount of moisture in their environment into *xerophytes*, *hydrophytes*, and *mesophytes*. Zoologists also classify animals according to their moisture habitat. In the case of crops, moisture in the soil is frequently the limiting factor. The humidity of the air is also an important element in variation production.

(4) *Temperature.*—This factor, a climatic one, is also a limiting element inasmuch as protoplasm, the physical basis of life, has its maximum and minimum limits of temperature. Within these extremes heat effects mainly the rate of growth, and each species has its own optimum temperature conditions. Certain butterflies assume dimorphic forms to correspond with the season of the year.

(5) *Chemical Agents.*—Inasmuch as growth and movement are the result of the metabolic activity of protoplasm, it is plain that anything that interferes with this activity will produce variations. Especially is this true of various chemical agents when they are excluded from or allowed free access to protoplasm. Oxygen, for example, is essential to the life of protoplasm, but on the other hand there are many kinds of poisons—catalytic, toxic and others,—and secretions that modify the activities of living cells and tissues.

(6) *Light.*—Light influences living matter in producing chemical effects, in promoting or hindering functional activity, and in its tropic effects. The coloring of leaves, fruit and skin is the result of the chemical action of light; the retardation of growth of stems of plants is the effect of light; and the direction of growth or movement is a tropic effect.

(7) *Contact.*—Animals and plants respond to the stimulus of contact. The effect may be observed directly upon protoplasm itself, and indirectly in movement.

(8) *Gravity.*—In plants the influence of gravity is very evident in determining the direction of growth of the chief members. Among the lower animals too the geotropic influence is often quite marked.

(9) *Use and Disuse.*—That organs develop with use and deteriorate with disuse are facts well known, and variations of considerable size undoubtedly arise by this process. The difference between the expert and the common man is largely due to the use or disuse of organs and faculties.

Chapter 4—DISTRIBUTION OF ANIMALS AND PLANTS

(a)—De Candolle's Law

This law relates to adaptation to climate. Through the action of Natural Selection, plants have become gradually adapted to the climate in which they live, and ill adapted to climates north or south. Just how far modifications may be due to heredity is not clear. After making a careful study of this matter De Candolle "concluded that native forms are not hardy when taken one hundred miles north or south of their source."

(b)—Laws of Distribution of Animals

"The laws governing the distribution of animals are reducible to three very simple propositions. Every species of animal is found in every part of the earth having conditions suitable for its maintenance, unless:—

(a) Its individuals have been unable to reach this region, through barriers of some sort; or

(b) Having reached it, the species is unable to maintain itself, through lack of capacity for adaptation, through severity of competition with other forms, or through destructive conditions of environment; or

(c) Having entered and maintained itself, it has become so altered in the process of adaptation (or as a result of other processes) as to become a species distinct from the original type."

Chapter 5—ORIGIN OF DOMESTICATED ANIMALS AND CULTIVATED PLANTS

(a)—Origin of Domesticated Animals

Dog.—It is probable that the dog was the first animal to be domesticated. Its origin is not definitely known, but as the wolf, the jackal, and the fox are close relatives it is believed he may have arisen from one or more of these wild forms. By man's selection the numerous breeds of to-day have gradually been developed.

Horse.—The wild form of the horse is also unknown. He probably developed in the plains of Central Asia and was domesticated by some tribe of the district. The ancestry of of the horse has been determined from fossil bones found in the rocks of Western United States. These remains show its evolution from a small five-toed animal the size of a rabbit up though a larger three-toed form to the still larger one-toed form like that of to-day.

Since his domestication the various breeds of horses have beeen developed by the selection of those forms that suited the fancy of man. Modern horses are no doubt larger, stronger and swifter than their undomesticated ancestors.

Cattle.—The origin of the chief breeds of European and American cattle is rather uncertain, but it is now believed that they are descended from two original types—the *Bos primigenius* or aurochs, and the *Bos sondaicus*, the former being large and powerful and long horned, the latter being much smaller and short horned. It is probable that the domestication of cattle began in Asia in pre-historic times. and that these numerous breeds were taken to Europe during the great migrations.

From the aurochs, it is believed, are descended the Holstein, Hereford, Scotch Highland, white cattle of Chartley, long-haired Hungarian and spotted Swiss breeds; and from the *Bos sondaicus* the Channel and the Brown Swiss breeds. The Shorthorns and the Ayrshires are probably a mixture of the two original types.

The Indian or Hindu humped cattle had their origin in a wild form (*Bos indicus*) that still inhabits the hilly country on the slopes of the Himalayas.

With the exception of the Holstein-Friesian, the French-Canadian and the Brown Swiss, American breeds of cattle are of British origin. For several hundred years

Great Britain has been noted as a great cattle-producing country, and the breeds, as we know them, were the result of isolation, selection and in—and line-breeding. Devonshire produced the Devon breed, Norfolk and Suffolk the Red Polled, Hertfordshire the Longhorns and Hertfords, Durham the Shorthorns, Scotland the Ayrshire, Galloway and Aberdeenshire-Angus, and the Channel Islands the Jersey, Guernsey and Alderney.

Famous Breeders.—Robert Bakewell (1726–1795) of Dishley Hall Leicestershire, may be considered the foremost stock breeder in a land of great breeders. His genius and skill were shown in the establishment of the improved Leicester breed of sheep as it is now known, and in the improvement of the Longhorn cattle and the Shire horse. His methods were adopted by later breeders, first by Charles and Robert Colling and afterwards by the Booths, Thomas Bates and the Cruickshanks, in the development of the Shorthorn breed to a high degree of excellence.

The names of Benjamin Tomkins and John Price are inseparably connected with the development of the Hereford; and Hugh Watson and William McCombie with that of the Aberdeen-Angus breeds.

Sheep.—Domestication occurred so long ago that the wild form is not definitely known. A large number of wild forms still inhabit the mountain districts of Asia and to a less extent in Africa and America. Possibly the two classes of domestic sheep—the horned and the hornless—are descended from separate wild forms.

Goat.—Unlike the horse, ox and dog the domesticated goat has had its origin in a single form. Its remains are abundant in the early period of the Swiss lake dwellings. The Pasang or Grecian Ibex (*Capra hircus ægagrus*), still found in Western Asia, is probably the ancestor of the domesticated form which has been bred for several thousand years in the Mediterranean Basin.

Swine.—Wild species of pig are to be found in many countries, but the common pig is descended from two wild species, the European wild boar and the Indian wild boar. Domestication took place in China centuries before it occurred in Europe, and it is known that some of the Chinese pigs were brought to Europe and crossed with the native species.

Poultry.—The ancestor of the common hen is very likely the Jungle fowl of India or the Indian Game Fowl

(*Gallus bankiva*), still to be found wild in the forests of India.

This bird is of bantam size, with the color of the brown Leghorn, and with the single comb, *i.e.* with a high serrate ridge. The various breeds of modern poultry have arisen gradually by crossing and long-continued selection. The fancy breeds are Bantam and game birds; the American breeds—Plymouth Rocks, Wyandottes, Rhode Island Reds and Dominiques; the English breeds—Orpingtons and Dorkings; the French breeds—Houdans and Crevecoeurs; the Dutch breeds—Hamburgs and Red Caps; the Mediterranean breeds—Leghorns, Minorcas and Black Spanish; the Asiatic breeds—Brahmas, Cochins and Langshans.

Goose.—Six breeds of geese are recognized; viz: Toulouse (grey), Embden (white), African (gray), Chinese (brown and white), Egyptian (colored), and Wild or Canadian (grey). The two first are slightly modified descendants of the grey lag (*Anser anser L.*) of Northern Asia and Europe. The Chinese goose is also derived from a distinct species (*Cygnopsis cygnoides*). The African breed is believed to be derived from the Toulouse, Embden and Chinese. The Egyptian goose comes from a different species, (*Chenalopex aegyptiacus*). The Canadian goose is the domesticated American wild goose (*Branta canadensis*).

(b)—Origin of Common Cultivated Plants

Wheat.—Wheat has been under cultivation for thousands of years and exists now as four well defined races—*Common, Hard, Polish* and *Egyptian.* No wild forms of these races have been found. Besides the true wheats, two other related species are in cultivation—*spelt* and *emmer,* and it is probable that the true wheats have been evolved from ancestral forms much like spelt. Recently Aaronsohn, a botanist of Palestine, found a wild wheat growing over large areas in Palestine, approaching in its characters to spelt and emmer. It cross-pollinates and is very resistant to drought and disease. It is believed that this wild wheat is the ancestor of our cultivated forms.

According to Virgil (Georgics I, 197) the Romans applied selection to their cereals:

> "The chosen seed, through years and labor improved,
> Was seen to run back, unless yearly
> Man selected by hand the largest and fullest ears."

Apparently, however, little improvement in cereals was made until the beginning of the 19th century, when Le Couteur of the Island of Jersey, on the suggestion of Le Gasca of Madrid, began the selection of ears of wheat which differed from others in the same field. One of these new varieties was the "Bellevue de Talavera," still grown in England and France. Later on, Patrick Shireff, a Scotchman, (1819) developed several new varieties by selection, namely: "Mungoswell" wheat, "Hopetown" wheat and oats, "Pringle's" wheat, etc.

In 1857 Hallett began his "pedigree culture" of wheat, oats and barley, and introduced the *Victoria, Hunter, Original, Red* and *Golden Drop* wheats, the *Pedigree White Canadian* and *Pedigree Black Tartarian* oats, and the *Chevalier* barley.

The *centgener* method of W. M. Hayes, of Minnesota, with modifications in some cases, is one commonly employed in cereal breeding, and consists in the sowing of about one hundred seeds from each of the mother plants selected for their excellence from a very large number of plants grown from the best grains. Several of the best plants from the centgener group are reserved for seed, and the total of each centgener group is weighed to estimate the comparative value of each of the original mother plants.

This process is repeated for the third and fourth years and the most promising varieties are planted in small fields or multiplying plots. (See also page 155).

Barley.—Three distinct races—the *two-rowed*, the *four-rowed*, and the *six-rowed*—have been in cultivation for thousands of years. It is the opinion that these races arose from a wild two-rowed ancestor in prehistoric times.

Rice.—This plant has been cultivated for more than 5,000 years in China and India where selection of superior strains was practised.

Rye.—Several related species of rye grow wild in Europe and Asia, so that there seems to be no doubt of the origin of the cultivated form. Its domestication, however, is comparatively recent.

Oats.—The modern races of oats are not of great antiquity. Several closely related wild species are to be found and it is very probable that our cultivated forms had their origin from one or more of these wild species.

Corn or Maize.—Although Maize is an American and a comparatively recent plant, its origin is unknown. The Indians cultivated it before the arrival of Europeans. It is probable that it was evolved from one or more of the many maize-like plants that grow wild in semi-tropical America.

Bean.—De Candolle is of the opinion that the common kidney or haricot bean is of South American origin, and was introduced into Europe soon after the discovery of America. The Lima bean is also a native of South America. The common broad bean of Europe, however, had its origin in the Mediterranean and Caspian basins and was cultivated by the lake-dwellers and the ancient Egyptians.

Pea.—Both the Field and the Garden Pea originated in the Mediterranean basin, where the *wild* form grows. The former is of recent introduction, but the latter is much older, having been cultivated by the Greeks and Romans, and even by the lake-dwellers. It originated probably from the wild Field Pea.

Clovers.—These legumes are of comparatively recent introduction, perhaps in the 16th century. Wild forms grow in the Mediterranean basin.

Alfalfa.—Although this plant is of recent introduction into America, it was cultivated thousands of years ago by the Persians. It grows wild in Western Asia.

Potato.—Evidence points to the Andean slopes as the native home of the potato. It was taken to Europe by the Spaniards, and then by them also to the United States.

Turnip and Cabbage are of recent introduction, domestication ocurring in Northern Europe where wild forms still ex st.

Apple.—The apple was cultivated by the lake-dwellers, the Greeks and the Romans and other races in the Mediterranean basin, but the fruit was small and clustered like the crabs. The increase in size and flavor belongs to comparatively recent times. The main modern varieties have origginated as chance seedlings,—probably mutations in many cases.

Jean Van Mons (1765–1842), a Belgian, and Thomas Andrew Knight (1759–1838), an Englishman, may very properly be called the fathers of modern fruit culture, the former demonstrating the value of *selection*, and the latter the value of *crossing* in the improvemen of plants. In the development of American fruits it is a noteworthy fact that many standard varieties were "accidental seedlings or chance discoveries of valuable wild forms."

Plum.—Two common strains of plums exist—the European (*domestica*) and the American (*americana*), each originating from wild species of which there are many. The plum was early domesticated, probably in the Mediterranean basin.

Peach.—The peach probably originated in Persia or India, but its wild form is not definitely settled.

Grape.—The grape has been cultivated for several thousand years, and it is believed that most of the present varieties of the old world have descended from *Vitis vinifera*, which is probably indigenous to Asia. The American cultivated varieties are derived from three native species, *V. rotundifolia*, *V. labrusca*, and *V. vulpina* or *riparia*, or by crossing with *V. vinifera*. The California grape is the *vinifera* of Europe.

Citrous Fruits.—The orange, lemon, lime, grape-fruit and citron, although not known in the wild form, are believed to have originated in Eastern Asia. They have been cultivated for several thousand years.

Strawberry.—Although there are a dozen wild species only a few have ever been brought under cultivation. From Keen's Seedling first known about 1821 most of the modern strawberries have descended. This seedling was derived from the Old Pine class which again was derived from an American species, *Fragaria chiloensis*. In America most of the modern varieties have come from Hovey's Seedling, a derivation of Keen's Seedling.

Rose.—The numerous varieties of roses have been developed by hybridization and clonal or bud mutations of many wild roses. The beautiful Hybrid Teas have risen from four species—the *chinensis, gallica, centifolia and damascena.*

Sweet Pea.—The original of this cultivated plant came from Ceylon in the pink and white Painted Lady. From this, white, red, black scarlet, blue and yellow color mutations as well as mutations in form, size, and habit have arisen. Later extensive hybridization and selection have developed the large number of forms of the present day.

Boston Fern—This fern arose as a bud mutation from *Nephrolepis exaltata*. By continued bud mutations of the Boston fern, the numerous forms of the present day have arisen.

It has been noted that practically all of our cultivated plants and domesticated animals are of prehistoric origin. As man progressed in the arts of civilization he drew upon wild nature for contributions. When we reflect, then, how few the domestications have been in historic times, we are obliged to believe that prehistoric man maintained for long ages a high civilization when fine selective skill and patient labor transformed wild life into cultivated fruitfulness and domesticated use.

These prehistoric people must have had "among them their Darwins and Vilmorins,[1] their Gartons and Burbanks, with the one important difference—that these achieved immeasurably greater practical results than have as yet their modern successors" (Thomson and Geddes, *Evolution*)

The region which witnessed such transformations was probably the Mediterranean basin, extending from Portugal through Asia Minor and Persia to Korea. Prehistoric cultivation terraces in this district still show how extensive were the plantations in ancient times.

Chapter 6—REPRODUCTION OF ANIMALS AND PLANTS

(a)—Kinds of Reproduction

A few words with regards to the modes of reproduction among plants and animals. For convenience we may divide organisms into *unicellular* and *multicellular*. In both divisions we find two methods of reproduction, viz.: *asexual* and *sexual*. Asexual reproduction, *i.e.*, reproduction without special germ cells or fertilization, occurs in several ways.

(1) *By division*: bacteria, many algæ; protozoa, sponges, coelenterates, and some annelids.

(2) *By budding:* most plants and lower animals.

(3) *By the formation of spores: most unicellulars and many multicellulars.*

(1)—The Vilmorin family has been perhaps the foremost plant breeders in France for over 150 years. Louis (1816-1860) was the producer of the improved sugar beet, and the inventor of the centgener and line method of breeding. His son Henri succeeded to the business and made valuable contributions to the study of heredity, and to the breeding of wheat and pototoes. He died in 1889 and was succeeded by his son Philippe (1872-1917) who was one of the most noted scientists of France. He wrote valuable works on wheat, sugar beet, ginseng, tobacco, and flower-gardening. Maurice, uncle of Philippe, is specially interested in Horticulture on which he has written several books.

Sexual reproduction, *i.e.*, reproduction by means of special germ cells, occurs in some unicellulars, and in most multicellulars. Three methods may be recognized:

(1) *Heterogamy*, where the eggs from one parent are fertilized by sperms from another parent;

(2) *Autogamy*, or hemaphroditism, where the eggs are fertilized by sperms from the same parent; and

(3) *Parthenogenesis*, where eggs may develop without fertilization, as among Aphids, and in *Taraxacum*, *Hieracium*, *Antennaria*, *Alchemilla*, *etc.*

Sexual reproduction in both plants and animals came into operation after the asexual. In the former, up to the highest forms, asexual and sexual methods have continued side by side, but in the latter sexual reproduction whenever introduced has displaced the asexual, with a few exceptions. Parthenogenetic development is rather rare, as it occurs in a few groups only, while autogamy or hermaphroditism is quite common among the higher plants and many of the animal groups such as the sponges, flat-worms, certain parasitic crustaceans, bryozoans and tunicates. It seems to be associated with parasitism or with a sedentary mode of life.

The Germ Cells or Gametes

The female germ cell is called the egg-cell or *ovum*. It is bounded by a cell wall and contains food substances and a *nucleus*. The latter, however, is the essential part as it contains the *chromatin* substance. The size of the egg-cell varies very greatly—usually minute, but large in birds—according to the amount of the food substance or *yolk*. In the sponges, corals, star-fishes, worms and mammals the yolk is small in amount and is diffused evenly through the egg cell. In the Arthropods the yolk occupies a central position. In the Amphibians the yolk lies at one end, and in fishes, reptiles and birds the yolk occupies nearly the whole egg-cell. The distribution of the yolk determines the method of segmentation after ferilization. (See Fig. 22).

The male germ cell is called the sperm-cell or *spermatozoon* (in animals). It is very much smaller than the ovum and varies in shape in the different groups of animals. In mammals it consists of a small pointed head, composed almost entirely of nucleus, and a long vibratile tail for purposes of movement.

When a sperm-cell meets an egg-cell it penetrates the outer coating and enters its substance. The tail is absorbed and the head or nuclear part moves toward the egg-nucleus which at the same time moves to meet it. The two nuclei merge into one. but the chromosomes of each retain their identity. This mingling of the male and female nuclei constitutes *fertilization* and the initiation of cell-division of cleavage of the egg-cell. The fertilized egg-cell by repeated divisions forms the embryo.

(b)—Fertilization in Seed Plants

The nucleus of the pollen grain divides into two—the *generative* and the *tube* nuclei—about the time of shedding of the pollen. The generative cell again divides into two *male* nuclei or cells. After the pollen grain reaches the stigma the pollen tube is formed which penetrares the style to the micropyle of the ovule in the ovary, thence through the nucellus to the embryo sac. The two male cells are then discharged from the end of the tube on reaching the egg-cell, when *fertilization* with subsequent development of the embryo is effected; the other fuses with the endosperm- or fusion-nucleus near the centre of the embryo-sac and forms the endosperm surrounding the embryo (Fig. 4).

(c)—Fertilization in Animals

Fertilization in animals, as in plants, consists in the fusion of a male cell with an egg-cell. This fusion takes place among the higher animals within the body of the female; but in frogs and fishes the union occurs after the eggs are laid.

(d)—Development of the Embryo in Seed Plants

(a) In *Shepherd's Purse* as a type of the Dicotyledonous plants. By several transverse divisions the fertilized egg-cell becomes a thread-like body called the *pro-embryo*. The last cell divides into octants, the four terminal cells forming the stem and cotyledons, the four basal cells the hypocotyl. By periclinal walls the *dermatogen* (or primitive epidermis) layer is cut off, while the *periblem* (or primary cortex) and the *plerome* (or primary stele) are soon differentiated from the interior tissues by many divisions (Fig. 5).

(b) In *Arrow-head* as a type of the Monocotyledonous plants. By two transverse divisions a three-celled pro-embryo is formed, the terminal cell forming the cotyledon.

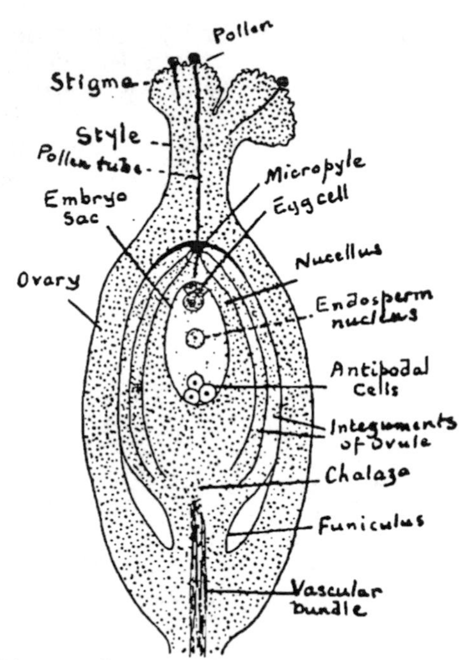

Fig. 1 - Diagram of an ovary of a plant with an ovule, showing their various parts, at time of fertilization.

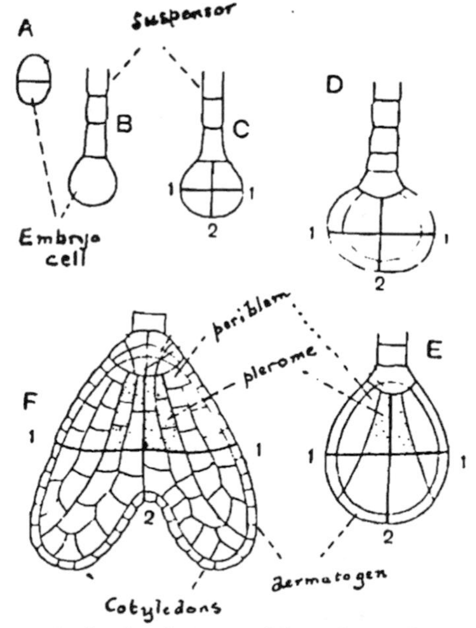

Fig. 5. -Stages in the development of the embryo of the Shepherd's Purse. Showing the octant section 1-1, and 2.

The middle cell by a series of divisions gives rise to the whole embryo, including the stem, hypocotyl and root tip. The stem tip arises laterally in a notch in the side of the embryo.

(e)—Development of the Embryo in Animals

The fertilized egg-cell divides rapidly by segmentation forming a mass of cells, but the nature of segmentation differs in different groups of animals according to the amount and distribution of yolk present. In mammals total segmentation occurs and the cells are nearly *equal* in size; in frogs and toads also *total* segmentation occurs but the cells are *unequal* on account of the large amount of yolk. In fishes, birds, reptiles, *partial* segmentation occurs on account of the excessive amount of yolk, and only a small disc-shaped mass of protoplasm lying on top of the yolk segments; in insects and crayfish there is also *partial* segmentation but the segmentating mass of protoplasm is *peripheral*, surrounding the central yolk. This stage of segmentation is called the *blastula;* the second stage, the *gastrula*, is formed by an inpushing or infolding of the blastoderm layer so that a body with two layers is produced —the *ectoderm* and the *endoderm* layers. Soon a third layer, the *mesoderm*, is formed between these two, chiefly from the endoderm. From these three layers all the tissues of the body are formed. From the ectoderm arise the sense organs, nervous system, and outer integument; from the endoderm the digestive portion of the alimentary tract, and from the mesoderm the skeleton, muscles, blood and lymph systems, reproductive organs and connective tissues. The manner of forming the animal body is different for the different groups, but in the vertebrates "the ectoderm forms first a shallow groove along the axis of the embryo, which becomes deeper by the more rapid growth of the sides, until the latter closes over and meets above to form a tube, the anterior end of which, by a further complication of folds, flexures and cell growth becomes the embryonic brain" (Abbott).

With regard to the segregation of the germ cells Weismann says that in one group of animals segregation takes place at the very beginning, and in the other group it does not occur until a later period. In *Ascaris* and in the Diptera among insects the fertilized ovum divides into two cells, one giving rise to the whole body (*soma*) and the other only to the germ cells lying in this body. In all other animals the primordial germ-cell appears later, after the first few divisions of the ovum, during or after the embryogenesis. Many facts now support the theory (*Evolution Theory, Vol. 1, pp. 410-411*). Jager and Messbaum had practically come to the same conclusion earlier for they had observed that at an early stage in the embryo the future reproductive cells are distinguishable from those forming the body. Researches since their time have shown the early origin of

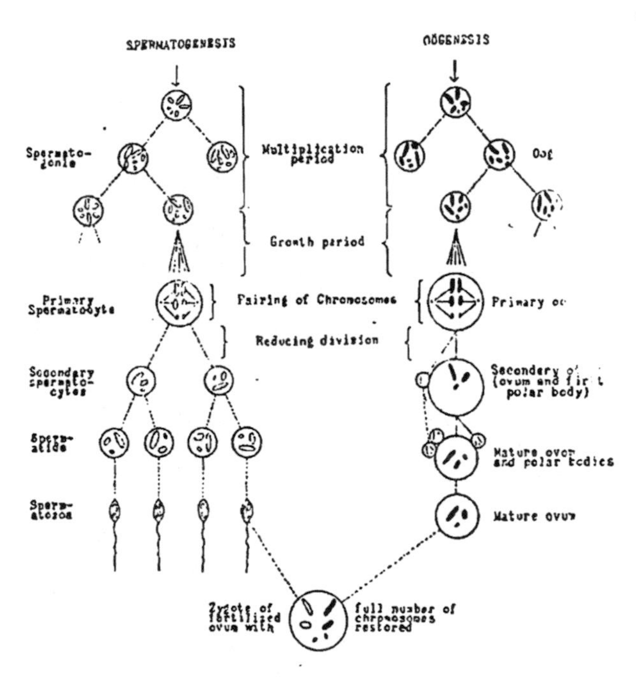

Fig. 6.—Diagram of gametogensis showing the parallel between maturation of the sperm cell and maturation of the ovum (After Guyer.).

germ-cells in leeches, threadworms, several insects, some crustaceans and spiders.

In the higher plants, however, soma plasm and germ plasm reveal no visible difference although in some false-graft hybrids Baur and Winkler found that one of the subepidermal layers is probably alone responsible for the sexual cells.

The point to be emphasized in development is the early setting apart of the primary germ cells before the body is formed. The body does not produce the germ cells, but merely houses and nourishes them.

During the formation of the body by the multiplication and differentiation of the body cells, the primary germ cells are relatively inactive, but at maturity they become active. By division *spermatogonia* and *ogonia* are formed, then *spermatocytes* and *oocyles*, and finally *spermotozoa* and *ova* (See Fig. 6).

Chapter 7—HEREDITY AND INHERITANCE

"Some are born great,—(*Heredity*).
Some achieve greatness,—(*Function*).
Some have greatness thrust upon them."—(*Environment*).
—*Shakespeare.*

Heredity has been well defined as "the genetic relation between successive generations," and inheritance, "all that the organism is or has to begin with, in virtue of its hereditary relation to parents and ancestors" (Thomson, *Heredity*).

The three factors that influence an organism are heredty, function, and environment.[1]

The heredity factor is often called *Nature*, and those of function and enviroment called *Nurture.*

The modifications produced by nurture are not transmissible, but they may be re-impressed on each generation. Every inheritance requires an appropriate nurture, if it is to realize itself in development. Nurture supplies the liberating stimuli necessary for the full expression of the inheritance. "A man's character, as well as his physique, is a function of nature and of nurture."

The determination of the laws of heredity is, perhaps, the most important of biological inquiries for the reason that if they were fully understood the problem of the mode of evolution of organic beings would be solved. But this

(1)—Dr. Chalmers Mitchell, in a recent volume entitled "Evolution and the War," discusses the factors operating in the making of nationality, and states that "the study of nationality is really a study of 'Kultur,'" i.e., the "whole set of forces, partly selective, partly directive, political, educational, social, environmental that go to the moulding of the national character, everything in fact that nurture can impose on plastic nature." He questions the opinions of Dr. Starr Jordan ("The Human Harvest"), Dr. Saleeby ("Parenthood and Race Culture"), Dr. J. A. Thomson and others regarding the effect of war upon the human stock; and chaffs those eugenists who claim that by selective agencies in breeding mental, moral and emotional qualities of a people can be modified.

Dr. Mitchell contends that "nurture is inconceivably more important than nature," and adduces many arguments in favor of the contention (Chap. V.).

matter of the genetic relation between successive generations is most complex, comprising, as it does, the consideration of many factors as yet but partially recognized and understood; however, much progress has been made in the study of heredity since Darwin's day, and several important laws have been discovered so that order is taking the place of chaos in some fields of inquiry.

From the standpoint also of the practical breeder of animals and plants a knowledge of the laws of heredity would be of great importance. The production of improved forms of our domestic animals and cultivated plants would be tremendously accelerated, and a corresponding increase in value of animal and plant productions would take place.

Many proverbs and phrases are current which show that the "common man" emphasizes the importance of heredity and environment. Some of these are "Like father, like son;" "Get a cat of a good kind;" "The fathers have eaten sour grapes, the children's teeth are set on edge;" "The corncrib cross;" "The breed goes in at the mouth;" "You cannot make a silk purse of a sow's ear;" One cannot gather grapes from thorns nor figs from thistles;" "The child is a chip of the old block;" "Wooden legs are not heritable, but wooden heads are;" "What's bred in the bone;" "Blood tells;" "Every good tree bringeth forth good fruit, but a corrupt tree bringeth forth evil fruit."

Methods of Investigating Heredity

Three lines of research have been conducted by students of heredity: (1) by studies of *statistics* of the progeny of crossing strains of plants and animals, after the manner of Galton and Pearson: *the Statistical Method;* (2) by *experimentation* after the manner of Mendel and his followers: *the Mendelian Method;* and (3) by the microscopic *study* of *the germ cells* and the behavior of the contained chromosomes, after the manner of Weismann and his followers: *the Biologic Method*.

(a)—The Statistical Study of Heredity

The statistical method of study of problems in heredity was first applied by Francis Galton. Later, important additions were made by Pearson and his associates, so that there is now a large number of reports based on statistical studies of various phases of evolution. The new science is called *Biometry or Biometrics*.

Biometry is of value in the determination of general laws relating to various phenomena of heredity in *populations* or in *races as a whole*. It furnishes information regarding the *trend* of evolution as influenced by operating factors, individually and collectively.

(b)—Galton's Law of Ancestral Inheritance (1889)

Galton formulated a law, based on data as to stature and other characteristics in man, and as to coat-color in Basset hounds. It may be stated as follows:

The parents together contribute one-half the total heritage; the four grandparents together one-fourth; the eight great-grandparents one-eighth, etc. This law, it should be borne in mind, is only true on an average for a large number of cases.

Karl Pearson has also investigated this law, and practically substantiates it. Mendelists, however, attempt to show that Galton's and Mendel's views are not yet in harmony. It should not be forgotten that one is a "statistical formula applicable to averages of successive generations breeding freely, and the other a physiological formula, applicable to particular sets of cases where parents with contrasted dominant and recessive characters are crossed, and their hybrid offspring are inbred" (Thomson).

Statistical methods for the study of inheritance are useful in cases which Mendelian analysis cannot solve. There are probably characters which are not inherited according to Mendel's laws; and there are some somatic characters determined by so large a number of factors, *e.g.*, stature, that their identification may be beyond the range of practical breeding.[1]

(c)—Galton's Law of Filial Regression

Galton and Pearson have also worked out another law of inheritance which may be stated as "the tendency to approximate to the mean or average of the stock." This is called the Law of Filial Regression. Galton's data dealt with the stature, eye-color and artistic faculty of about 150 families. Pearson's conclusions are given as follows: "Fathers of a given height have not sons all of a given height, but an array of sons of a mean height different from that of the father, and nearer to the mean height of sons in in general. Thus, take fathers of 72 inches, the mean

(1)—Other complex problems in heredity are the speed of race-horses, milk production in cattle, and the texture of wool when breeds are crossed.

height of their sons is 70.8 inches, or a *regression* towards the mean of the general population. On the other hand, fathers with a mean height of 66 inches give a group of sons of a mean height 68.3 inches, or a *progression* towards the mean of the general population of sons. The general result is a sensible stability of type and variation from generation to generation" (Pearson, *Grammar of Science*).

The following chart shows graphically the results obtained by Pearson in his investigation where more than a thousand pairs of fathers and sons were measured. The line AB joining the points in the chart is nearly straight but not horizontal,—in fact about half way to 45°.

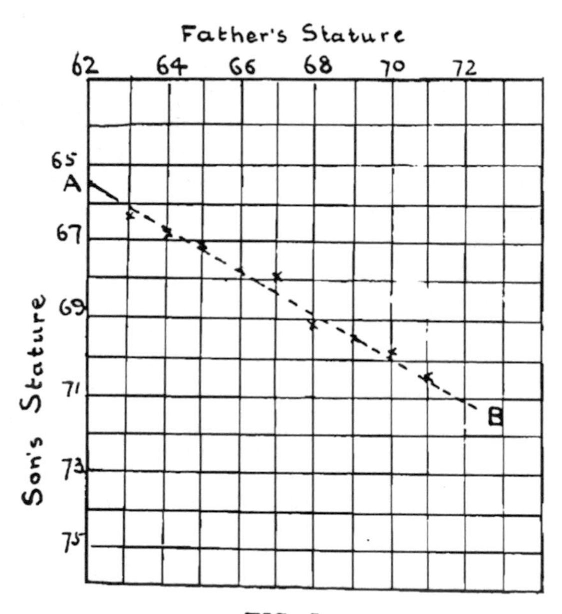

FIG. 7.
Chart to illustrate Galton's and Pearson's Law of Filial Regression (after Watson).

The interpretation is:—if the line were horizontal, *i.e.*, if the angle with the horizon be 0°, all the sons would be of same average height, and the inheritance would be zero (tan 0 = 0); if the line were inclined 45° all the sons would be of the same average height as the fathers and inheritance would be unity (tan 45° = 1). As the line is about 23° the average height of the son is about half way between the height of the father and the mean height of the race. This fraction is termed the *co-efficient of heredity*, and is expressed mathematically by the *tangent* of the *angle of inclination* from the horizontal.

This coefficient of heredity for stature in man (=.51) is high, "which indicates either that stature is very largely a matter of heredity or that environmental conditions usually tend to have similar effects on father and son." Studies, show, however, that the coefficient of heredity depends altogether on the nature of the case. Within a "pure line" the coefficient is zero; in a mixed population of sweet peas the coefficient is about unity. The coefficient of heredity for speed in horses is not known but would be valuable.

(d)—Value of Biometric Studies

It is plain that statistical results apply to *averages* and not to *individuals*. In heredity biometrics does not take into account the kinds of varieties but the gross total, and for this reason it does not help the geneticist, who desires to know what will definitely occur in any particular case.[1]

Castle says that Galton's Law works fairly well in cases of blended inheritance, but "as a useful generalization it is now pretty generally discredited. The reason is fairly obvious. It was an attempt to unify in classification things of unlike character viz., blending and Mendelian inheritance."

Biometrics has nevertheless proved useful in showing the falsity of many current beliefs. For example, it is a popular belief that in human matings "opposites attract each other." Pearson shows clearly that there is a decided tendency for "like to mate with like." He has found, moreover, that "tall women procreate faster than small women, also that dark-eyed people are more fertile than light-eyed.

Chapter 8— UNIT CHARACTERS[2]

As a result of the investigations of Mendel, DeVries and others, there has arisen the conception of *unit characters*, which enters largely into recent studies of heredity.

(1)—A geneticist has made the following facetious comparison between statistical and Mendelian methods of studying heredity. It is implied in the answer to the question: Why do white sheep eat more than black sheep?

(2)—Herbert Spencer has the honor of first putting forward the idea of organic units in connection with a theory of heredity. In the "Principles of Biology" (1864) he called these "physiological units." They circulate through the body and "in course of time visit all parts of the organism." Moreover, he assumed that the germ-cells are derived from the parent-body and that modifications impressed upon the body produce similar modifications on the offspring.

Each organism, according to DeVries, consists of thousands of elementary entities or *units* which combine to give it its form and functions. Similarity of species consists in the similarity of these elementary unit characters, while divergent species contain one or more different units. A unit may express itself in different organs, so the principle of units may help to explain the large number of correlations often discernible in plants. "The correlated external marks may be assumed to be the expressions of the same internal characters."

In an organism a number of characters may be distinguished capable of varying independently and of being isolated or followed separately in breeding experiments, each depending on a special transmissible germinal factor or factors. These characters are known as *unit characters*.

The idea of unit characters lies at the base of the modern theory of hybridism—Mendelism. "All the modes of expression of one unit must steadily keep together whenever the entire group of characters are thrown into one another in crossing."

Unit characters cannot be split or broken up, and when a unit is added or subtracted from a type a mutation is produced.

It sometimes happens that a character will disappear entirely in the course of time. Such a form is termed by DeVries a *retrograde variety* When an entirely new character appears it is termed a *Progressive Mutation.* DeVries considers this kind to have been the chief factor in evolution.

When a character appears that has been latent it is said to be *Degressive Mutation.*

"Varieties differ from elementary species in that they do not possess anything really new. They originate for the greater part in a negative way by the apparent loss of some quality; and rarely in a positive manner by acquiring a character already seen in allied species" (DeVries, *Species and Varieties.*)

(a)—Correlation of Characters

It is frequently observed that a particular color in a flower is associated with a particular taste or color in the fruit or seed, or a particular color with a particular form.

Judges and students of live stock have long been familiar with the fact that certain characters always go together as if they were produced by the same cause and were expressions of the same character. One must be careful, however, to distinguish between correlation and mere association of characters. Correlated characters move together —for example, short-beaked pigeons have small feet, hairless dogs have imperfect teeth, and blue-eyed tomcats are deaf.

It is not always easy to determine to what extent characters really move together. It is not safe to trust to mere impressions of a correlation; not until actual measurements and calculations are made can one be sure of the movement together of characters.

Correlations often exist in plants between botanical marks and the breeding qualities. Nilsson's work at Svalof, in Sweden, has shown the value of such correlations. Take for example, the selection of Primus barley, of different strains of peas, oats, and clovers, etc. This principle of correlation of characters is being followed up by many experimenters.

Burbank[1] uses intuitively this principle in his large selections, for he makes the majority of them while the plants are in the seedling stage. He can "predict one quality or one function from the study of others."

This correlation of characters brings forward again the idea of unit characters which forms the basis of Mendelism. The idea of correlation forces on us the assumption that the unit may express itself in many ways—in the leaf, seed, stem or tissue. "The correlated external marks may be but the expression of the same internal character."

The recent work of Morgan with *Drosophila* brings out clearly the fact that "every change in the germ-plasm (variation) affects not one but a large number of characters; and conversely, every visible character is the result of the concurrent action of a large number of factor-differences or variations.

(b)—Coefficient of Correlation

The ratio of correlation between separate characters is called the Coefficient of Correlation. When the coefficient is 1 the two characters involved are perfectly correlated, when 0 they are indifferent, and when -1 they are mutually exclusive.

When inter-relations are expressed in the form of a coefficient or ratio many errors of judgment, and truths as

(1)—Luther Burbank (1849-) lives at Santa Rosa, Cal., where he conducts his experiments on the breeding of new plants. Some of his productions are the Burbank potato, Burbank plum, plumcot, seedless prune, royal walnut, spineless cactus, etc.

well, are likely to be revealed. The following problems at first sight suggest certain correlations which may be far from the truth:

(1) What is the correlation between length and weight in ears of corn?
(2) Between white cats and deafness?
(3) Between black pigs and cholera?

Yule's formula is generally used in problems of correlation involving presence or absence but not degree, as in problems (2) and (3) above. M and N are the characters,

	M present	M absent
N present	a	b
N absent......	c	d

$$r = \frac{ad - bc}{ad + bc}$$

When the characters are present in varying degree as in problem (1) above, another formula is employed:—

$$r = \frac{\sum D_L D_W}{n S_L S_W}$$

D_L = deviation from mean length (or size etc.) of ears.
D_W = deviation from mean weight of ears.
S_L = standard deviation of length of ears.
S_W = standard deviation of weight of ears.
(See Davenport, *Principles of Breeding*).

Chapter 9—THE TRANSMISSION OF CHARACTERS

(a)—Darwin's Theory of Pangenesis (1869)

Several theories have been proposed to account for the transmission of *modifications*. Darwin, in his *Variation of Animals and Plants under Domestication*, suggested that very minute particles, called gemmules, are given off from all the cells of the body, which multiply by fission, and are ultimately carried to and aggregated in the germ cells. In development the gemmules unite with others like themselves, and grow into cells like those from which they were originally given off. This theory has been shown untenable by Galton and others; besides it is based on the erroneous idea that the body makes the germ cells.

(b)—DeVries' Theory of Intracellular Pangenesis or The Pangen Theory (1889)

This theory may be summarized as follows:

Organisms are built up of unit characters which are independently variable and heritable. They are represented in the germ plasm of the nucleus by definite bodies (pangens) which constitute the chromosomes. These pangens multiply in the idioplasm of the nucleus. Some of them migrate into the surrounding cytoplasm, where they become active and give it a particular character. A small number of pangens always remain in the nucleus, and is handed on from cell to cell by nuclear division. Into each cell, as it is formed, a fresh migration of pangens occurs.

"It will be observed that DeVries drops Darwin's idea of migration of the gemmules from the organism into the germ plasm, and starts with these gemmules as permanent constituents of the germ plasm."

(c)—Weismann[1] and Weismannism

Weismann's main contribution to evolution literature is his *Germ Plasm Theory*. This is primarily a theory of heredity, and only when it is considered in connection with other related matters does it become the theory of evolution called Weismannism (Fig. 9).

Weismann (1834–1914) argued that as all higher animals and plants have originated from a single cell, a fertilized egg, this minute structure "must contain all the hereditary qualities, since it is the only material substance that passes from one generation to another. This hereditary substance is the germ plasm, and Weismann's special theory is called *the Continuity of the Germ Plasm*.[2] The fertilized egg-cell is the result of fusion of two germ-cells, the larger, the female egg-cell or ovum, and the smaller, the male sperm-cell. Since the offspring often contains characteristics of both parents it is also evident that each germ-cell must be the bearer of special hereditary qualities. The body or *soma*, according to this idea, is merely "an offshoot—a house built up out of a part of the substance of the original germ-cell to shelter it until it decays, and the germ-cell is transmitted to another house."

(1)—Weismann studied medicine at Göttingen, and for over fifty years was teacher and professor of Zoology in Freiburg University. He was afflicted with an eye trouble which interfered with his work for many years.

(2)—This theory was first formulated by Francis Galton in 1875 in his "Stirp" theory of heredity.

"Germ Plasm has, therefore, had unbroken continuity from the beginning of life, and owing to its impressionable nature it has inherited an organization of great complexity."

"In development a part of the germ plasm (*i.e.*, the essential germinal material) contained in the parent egg cell is not used up in the construction of the body of the offspring but is reserved unchanged for the formation of the germ-cells of the following generation" (Weismann).

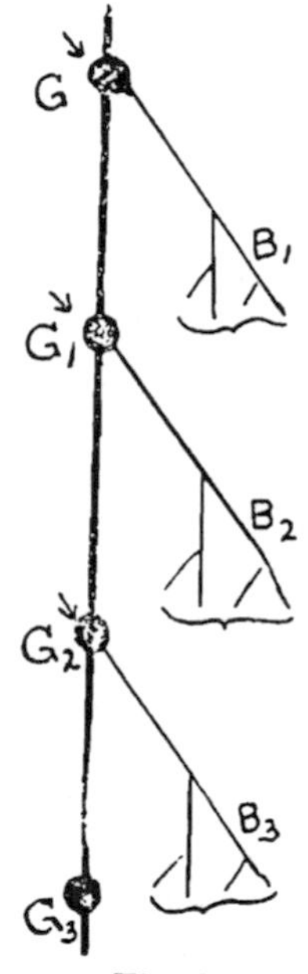

Fig. 8.
Diagram illustrating idea of continuity of germ plasm. (After Wilson).

The relationship between germ plasm and body—or soma-plasm is shown in the following diagram:—

G is the fertilized ovum which divides into body cells B_1 and the germ-cells G_1. G_2, G_3, G_4, G_5, are the lineage or chain of germ cells and B_2, B_3, B_4, B_5, are the lineage of body-cells.

In development a part of the germ plasm contained in the parent egg-cell (G) is not used up in the construction of the body (B_1) of the offspring, but is reserved unchanged for the formation of the germ-cells of the following generation.

The difference between the old conception and Weismann's conception of the relation between successive generations is naively put by Samuel Butler: "It is not to say that the hen produces another hen through the medium of an egg, but to say that a hen is merely an egg's way of producing another egg."

Variations, according to the Germ Plasm theory, are the result of the union of the male and female germinal elements. The mixture of these plasms (*amphimixis*) in fertilization gives great possibilities of variations arising from the different combinations and permutations of the vital units within the germ plasm. In single-celled organisms Weismann states that the environment produces variations directly.

Weismann, moreover, denies that acquired characters are transmitted from parent to offspring. His argument is as follows: "Acquired characters affect the body cells, and the latter are simply a vehicle for the germinal elements, which are the only things concerned in the transmission of hereditary qualities. Inheritance, therefore, must come through alterations in the germ plasm, and not directly through changes in the body-cells "[1] Besides, Weismann adopts and extends the principle of Natural Selection of Darwin by the conception of *Germinal Selection*. He represents a "struggle among the determining elements of the germ-cell's organization. It is at least conceivable that the stronger "determinants," *i.e.*, the particles embodying the

Fig. 9.—AUGUST WEISMANN

rudiments of certain qualities, will make more of the food supply than those which are weaker, and that a selective process will ensue" (Thomson).

Panmixia.—Weismann attempted to explain how useless organs degenerate under the operation of Natural Selection by his theory of *Panmixia*. When under new conditions certain organs become useless natural selection ceases to operate, and the individual will no longer be at a disadvantage and therefore will survive. The crossing of

(1)—The idea of Germinal Continuity was suggested also by Owen (1849), Haeckel (1874), Rauber (1879), Galton, Jager, Brooks (1876), and Nussbaum.

individuals with a more or less useless organ will result in the production of progeny that is below the general level of efficiency, and the organ will appear degenerate.

General mixing of this nature Weismann called *Panmixia.*

Objections to Weismann's Theory.—The main objecttions come from botanists who find in plants the common phenomenon of *regeneration* of the whole organism from parts which do not contain the reproductive organs. In plants like *begonia* and *marchantia* the vegetative cell must contain the same *initials* as the germ cell, and there is no difference in the hereditary characters.

1. The hereditary characters are not confined to the germ-cells, *e.g.*, in plant regeneration by cuttings, grafts, etc. Weismann supposes that in the higher plants germ plasm is contained in a great many cells, in a latent state, and only becomes active according to the influences bearing on it.
2. The chromosomes are not definite organs of the cells, as they disappear in the anaphase stage of cell division (see Jost).
3. The chromosomes do not constitute the whole of the nucleus, and perhaps the hereditary capacity lies in some other nuclear substance (See Jost). "In *Lilium* it has been demonstrated that when fusion occurs there is no cytoplasm whatever investing the male nucleus."
4. In certain hybrids, *i.e.*, *Cytisus adami* (*C. laburnum x C. purpureus*) initials are sometimes lost (see Jost and Baur). *Cytisus adami* is a sterile hybrid and has most organs intermediate in character. It splits in a vegetative way, reverting to its parents.
5. A possibility of migration of idioplasm from one cell to another.
6. The germ plasm must be preserved along with the soma plasm in plants when not in flower, else where does the germ plasm reside in the sporophyte?
7. The germ plasm in the sex elements of flowers may be directly exposed to external influences (Henslow, *The Origin of Floral Structures*).
8. The non-transmissibility of acquired characters is not a corollary of the continuity of the germ plasm.
9. Asexually multiplied plants show variations and these variations may be transmitted from generation to generation (Bailey).

Weismann's Theory regarding the Constitution of Germ Plasm.—"The physical basis of inheritance—the germ plasm—lies in the *chromatin* of the nucleus of the germ cell, which takes the form of a definite number of chromosomes or *idants*.

"The chromosomes consist of *ids*, each of which contains a complete inheritance.

"Each id consists of numerous primary constituents or *determinants*.

"A determinant is usually a group of *biophors*, the minutest vital units.

'The biophor is an integrate of numerous chemical molecules (Compare with Morgan's views).

Chapter 10—CONGENITAL AND ACQUIRED CHARACTERS

Two facts stand out prominently in the study of successive generations of individuals in both plants and animals:

(1) The strong resemblance or the persistence of like characters from generation to generation, the idea being often expressed by the phrase "like begets like;" and

(2) the occurrence of variations among individuals even of the same parentage, expressed by the phrase "like begets unlike."

These conclusions hold true whether the plant or animal is produced asexually or sexually. There are, it is true, differences among organisms in degree of variability, but the fact of universal variability remains.

In the discussion of characters that may be transmitted two kinds are generally recognized, viz.: *congenital*, or more correctly *blastogenic. i.e.*, those characters "that have arisen through causes affecting the germ plasm directly; and *acquired* or *somatogenic*, those that affect the body during its development.' It is conceded that congenital or blastogenic characters can be and are transmitted, but it has long been disputed whether acquired or somatogenic characters can be transmitted (Neo-Darwinian and Neo-Lamarckian discussions). In other words, it is not conceded that "structural changes in the body induced by changes in functional or environmental influence will specifically affect the reproductive cells; that these will, if they develop, reproduce the modification acquired by the parent or parents."

The weight of evidence seems to be against the theory of transmission of acquired characters.

Summarizing, we may arrange characters as follows:

An individual is made up of
- A. *Congenital Characters*
  - (a). *Inherited* (growing under the stimulus of function and environment.
  - (b) Variations.
- B. *Acquired Characters* obtained
  - (a) by action of environment,
  - (b) by function.

(a)—Inheritance of Acquired Characters or Somatic Modifications

It will be recalled that Lamarck's theory of evolution involved the inheritance of the acquired characters obtained through the effects of use and disuse of parts. Darwin's Theory of Natural Selection also implies the heritability of useful variations. But this proposition has been challenged by many recent investigators, the chief being Weismann. E. Davenport (*Principles of Breeding*) and others assert that the term "acquired characters" should not be used, inasmuch as the modifications referred to are "differences in degree not in kind," and, therefore, are not *new* characters nor *acquired characters*. In the discussions on this question, as a rule, zoologists are opposed to the idea of transmission of acquired characters, and many botanists are in favor of it. The two camps have been called the Neo-Darwinian and the Neo-Lamarckian respectively.

The distinguishing tenets of Neo-Darwinism are: Variation is of sexual or internal origin, and acquired characters or somatic modifications are not hereditary. Those of Neo-Lamarckism are: External causes or the environment are directly responsible for much variation, and acquired characters are often hereditary.

Everyone knows that modifications are acquired by the individual during its life-time, under the influences of the conditions to which it is exposed. Living things must in many cases adjust themselves to the changing conditions of their surroundings. Usually the penalty of non-adjustment is death.

Acquired characters or somatic modifications may be arranged into four groups:

(1) *Mutilations.*

(2) *Diseases.*

(3) *Adaptations to external conditions.*

(4) *Functional adaptations.*

With regard to mutilations there is no trustworthy evidence that such are heritable.[1] It is difficult, moreover, to believe that the modifications made on our bodies by accidents, by disease, by exposure to environment, or by the functional activity of parts, for such are occurring every day, are heritable. Brown-Sequard's experiments on guinea pigs, in which he produced epilepsy by the section of certain large nerves and epileptic symptoms appeared in the offspring of a few of the animals, have been frequently quoted in proof of the transmissibility of acquired characters. Later investigators, however, failed to confirm Brown-Sequard's experiments; besides, his results are capable of another interpretation (Consult Thomson's *Heredity*).

Some experiments have been conducted in transplanting plants and animals to a new environment, when marked changes of habit occurred. But here again no conclusions of a definite nature can be drawn.

In the discussion of this subject the real question at issue is: Does a structural change of this kind in a part of the body so influence the germ plasm that the offspring will show the same modification? No one doubts that environmental influences may have an indirect influence on the offspring, but this is not to the point.

It must be confessed that much of the discussion as to the heritability of acquired characters has arisen through a misunderstanding of the exact nature of acquired characters and in applying loose methods of reasoning.[2] Thomson ably summarizes these and similar misunderstandings after the following manner:

1. *Interpretations are not necessarily facts.*—Many facts of nature, such as the disappearance of the legs of snakes, the hardening of the hoofs, the long neck of the giraffe, etc., are explained as due to the action of use and disuse, but these interpretations may be erroneous.

2. *Begging the question.*—It is often assumed that certain characters such as short-sightedness and rheumatism are modifications, and transmissible. But may these not be congenital characters ?

(1)—For example, the right of circumcision and the docking of lambs' tails have been practised for hundreds of generations.

(2)—Lamarckians object to the practice of Weismannists of calling acquired characters congenital as soon as they are shown to be heritable.

3. *Mistaking the reappearance of a modification for the transmission of a modification.*—For example: Nageli brought Alpine hawkweeds to Munich, and as a result of the new conditions they assumed new habits, which their descendants also possessed. These modifications were impressed upon each generation by the new conditions, but when these plants were returned to their Alpine habitat they assumed their Alpine characters.

4. In the case of microbic diseases, *mistaking reinfecttion for transmission.* Simply because a parent diseased with tuberculosis or syphilis may have offspring affected with the same disease does not make these diseases heritable. The children most probably become affected before birth during the gestation period.

5. *Changes in the germ cells along with changes in the body are not relevant.* That alcoholism runs in families may be perfectly true; the probable explanation is that germ cells, as well as body cells, are poisoned, and the offspring may show similar peculiarities as the parent. It may be, however, that a deficiency of control of alcohol is inherited.

6. *Failure to distinguish between inheritance of a particular modification and that of indirect results, or of correlated changes of that modification.* The blacksmith, for example, acquires a strong right arm through use, and his occupation is one that contributes to the healthy nutrition of every organ of the body. It is quite possible that the children of blacksmiths have stronger right arms than the children of parents engaged in sedentary occupations—on account of the indirect or secondary effect of the better nutrition of the whole body, including the germ cells.

The old Hebrew proverb: "the fathers have eaten sour grapes and the children's teeth are set on edge" (Ezekiel), represents the belief in inheritance of modifications. As Thomson remarks, one would have to enquire carefully, however, whether the children had not been in the vineyard too, before coming to a conclusion that the acquired character in this case had not been transmitted.

7. *Appealing to data from fewer than three generations.* This is quite a common mistake in discussions of this nature. Modifications may occur for one or two generations, but disappear in later generations.

8. *Transmission in unicellular organisms is not to the point.* As they have no "body" distinct from the germ plasm, the term *acquired characters* does not apply (Thomson, *Heredity*).

Tower's experiments with Colorado potato beetles seem to show that environmental influences may affect the germ plasm to such an extent that the progeny will show permanent variations. He subjected some of the beetles to changed conditions of heat and cold and moisture, when the reproductive organs were at a certain stage of development. The progeny were decidedly paler than their parents, but were fully as healthy. Moreover, these pale beetles when mated produced offspring like themselves, and so on for subsequent generations.

Some recent experiments with alcoholized guinea-pigs reveal the fact that the germ plasm is affected, for the majority of the offspring of such guinea-pigs are defective in many particulars.[1]

(b)—Telegony

Telegony is the supposed influence of a previous sire on offspring subsequently borne by the same female to a different sire. There is a widespread belief in Telegony, and many examples can be cited in its support. The case of Lord Morton's mare, cited by Darwin, is a classic example (*Animals and Plants Under Domestication*, and Thomson, *Heredity*). A nearly pure-bred Arabian chestnut mare bore a hybrid to a quagga stallion. Subsequently she bore two colts, a female and a male, to a black Arabian stallion These colts were dun-colored and were striped on the legs, one having in addition stripes on the body and neck. They had also quagga manes, the hair being short, stiff and upright. Lately, Prof. Ewart, by repeating the experiment, (the Penycuick Experiment) has shown that there is no proof of any influence of a previous impregnation.

Prof. Ewart points out that several marks of the ancestral forest type, such as the yellow dun color, the dorsal band, the zebra-like bars on the legs and often the faint stripes on the face, neck and withers, are quite common among Arabian crosses.

In Lord Morton's crosses "the bars on the legs were more marked on the hybrid, on the filly, and on the colt than on the quagga."

More recent experiments by Baron de Parana and the U.S. Government (Rommel) confirm Prof. Ewart's conclusions as to the untenability of the theory of telegony. (Read Chapter 36, *Genetics in Relation to Agriculture* by Babcock and Clausen).

(1)—Are Somatic variations ever inherited according to Mendelism? (See Emerson, Amer. Nat. 48, 1914).

Transplanting the Ovaries.—The results of the experiments of Castle and Phillips with guinea-pigs support the views of Weismann. The ovaries of a young black guinea-pig were transplanted to a young white female whose ovaries had been previously removed. This white guinea-pig was later mated to a white male. The progeny in three litters were all black.

Are the effects of training hereditary? Account for the improvement in the speed of trotting horses in the last hundred years from a mile in 2:48½ by Barton in 1810 to a mile in 1: 58 by Uhlan in 1913, and the records of college athletics.

Evidence[1] is accumulating to prove that alcohol is both a germinal and a foetal poison, and is in some degree responsible for the large percentage of unhealthy and imbecile children born of alcoholic parents, and for the large percentage of still births and abortions. One must be careful, however, not to fall into the old fallacy, "Post hoc ergo propter hoc." Alcoholism may possibly be only a symptom of some neurotic taint, expecially feeble-mindedness. It may be that the defects in the children are not really due to the effects of alcohol but to the fact that the parents were degenerate to begin with.

There is no doubt, however, that alcohol has a great affinity for the reproductive glands and in experiments conducted in alcoholized mammals the amount in these glands was about 3/5 of that in the blood (See pages 169-170).

(c)—Maternal Impressions

It has always been a very common belief that certain "vivid sense-impressions of a pregnant mother may so affect the unborn offspring that structural changes result, which have some correspondence with the maternal experience." Jacob set up peeled wands to get striped cattle in his efforts to outwit Laban (Genesis XXX. 31–43). Birth marks are commonly ascribed to prenatal disturbances.

When it is borne in mind that there is no direct means of transmission from mother to child, for no nerve or blood-vessel passes through the placenta, it is hard to imagine that "impressions" are produced in this way. The child derives its nourishment not directly but indirectly by a process of *soakage* or osmosis from the mother's blood.

(1)—Nicloux, Forel, Bezzola, Goddard, Mjoen, Stockard, Pearson, are some of the investigators whose publications should be read.

Again, it is generally believed by embryologists that physical defects or errors of development are due to intra-embryonal causes, and that most of them occur in the first two or three weeks. "Maternal impressions," on the other hand, are in the great majority of cases referred to a period after the fourth or fifth month. Usually explanations are found after the events, and these are put in the place of *causation* instead of *coincidence.*

While there is little or no scientific evidence in support of the theory of maternal impressions as usually understood, it is scientifically correct to assume that the prenatal condition of the mother may have an indirect influence on the health of the child. Beside abnormal nutrition, heredity, however, is largely responsible for the transmission of peculiarities to the child. "A knowledge of the pedigree of Laban's cattle would undoubtedly explain where the stripes came from."

Weismann went too far, we believe, when he tried to show the impossibility of the inheritance of acquired characters. The soma plasm and the germ plasm are not autonomous, physiologically at any rate. Recent studies go to show that *hormones, chalones,* or *internal secretions* of certain glandular bodies and tissues in animals play an important part in growth. The secretions of the so-called ductless glands, thyroid, pituitary, supra-renal, and spleen, and of the interstitial tissues of the testes and ovaries, when liberated into the blood, are carried to various parts of the body and influence their growth. Some of the hormones appear to link up the activities of the somatic with the germinal substances (See Parker, *Biology and Social Problems*).

MacDougal injected dilute solutions of zinc sulphate and other substances into the ovaries of *Raimannia* immediately before fertilization. The seeds that set produced plants different from the mother plant, and the difference was transmitted to succeeding generations.

Parallel Induction.—Many cases might be cited where both the germ plasm and the soma plasm are affected at the same time by an external stimulus. Prof. Gage fed poultry with the aniline dye Sudan III with the result that the dye not only appeared in the fat tissues but also in the eggs and in the fat tissues of the chicks. Sitkowski fed the larvæ of the clothes moth with the same dye, and the moths laid colored eggs which produced larvæ tinged with the dye.

These are examples of *parallel induction.*

Summary

In summing up the evidence as to the heritability or non-heritability of acquired characters, we may state the present position as follows:

1. The heritable characters are transmitted through the germ plasm (sperms and eggs) which is continuous from generation to generation and which is not produced by the body (soma). (See page 54).
2. No mechanism exists, as far as is known, whereby a structural change in a part of the body influences the germ plasm in such a way that the offspring shows the same modification.
3. All so-called instances of inherited acquired chararacters are based either on a mis-understanding of the exact nature of such characters or on loose methods of reasoning. None have satisfactorily met the test of rigid experimental proof (See page 60).
4. The persistence of the chromosomes and Mendelian factors is wholly opposed to the idea that the body influences specifically the germ plasm. (See chapter 16).

It is evident that the subject has not been an easy one to settle, when one rembers that such eminent men as Lamarck, Herbert Spencer, Haeckel, Hertwig, Cope, Hyatt and Sir W. Turner have been advocates of the theory of the inheritance of acquired characters; while opposed to the theory have been Darwin, Wallace, Galton, Huxley, Ray Lankester, Weismann and His.

Chapter 11—DE VRIES' MUTATION THEORY (1901)

We have already noted (page 20) that one of the objections to the full acceptance of Darwin's *Theory of Natural Selection* as an interpretation of the mode of evolution was the great length of time it demanded for the development of new species, not to speak of genera, families and classes, of plants and animals. The theory practically removed the problem of evolution beyond the range of experimental investigation or proof. Such an unscientific attitude did not appeal to many critics, and a few resolved to find out by observation and experiment if new forms might not possibly originate in a much shorter time than that demanded by Darwin's theory.

Already investigators had evidence that new forms had occasionally arisen suddenly as sports, and St. Hilaire, a co-tempory of Lamarck, had suggested that developments might occur suddenly by leaps. Both Huxley and Galton believed that nature makes jumps now and then, while Bateson had accumulated a large number of instances of discontinuous variations, and was convinced that such variations were common. The importance of discontinuous variations as factors in evolution was clearly shown by Hugo DeVries in his "Mutations Theory," published in 1901. (Fig. 10).

According to this theory species have arisen after the manner of *spontaneous* or *discontinuous variations* in contra-

Fig. 10.—HUGO DeVRIES

distinction to their origin by the selection of *fluctuating* or *continuous variations*, as proposed by Darwin.

It is now recognized that Natural Selection does not explain the origin of species or that of adaptations, but rather the persistence of adaptations and the elimination of the unfit. DeVries' *Mutation Theory* attempts to account for the origin of specific characters.

"Mutations arise suddenly and without any obvious cause; they increase and multiply because the new characters are inherited. Mutations are not necessarily large, many are smaller than the differences between extreme fluctuating variants. DeVries' species are termed *Elementary Species.* The Linnæan species of the systematist are artificial groups, and not those presented by nature. Such natural species do arise in the garden and in agricultural practice, as shown by DeVries. As presented by DeVries,[1] the Mutation Theory is not an alternative theory to Natural Selection, but a supplementary hypothesis. "The special problem which the Mutation Theory seeks to explain is the manifold diversity of specific forms."

According to the Mutation Theory the struggle for life occurs among species, as well as among individuals. Natural Selection of species brings about a survival of the fittest species, eliminating some and protecting others. In this way Natural Selection "guides the development of the animal and vegetable kingdom."

(a)—**DeVries' Experiments**

(Consult DeVries' *Species and Varieties*)

The importance of DeVries' experiments lies in the application of the experimental method to the question of the origin of specific characters. "The solution of this problem," he said, "must be sought among the facts themselves." He actually observed the origin of new plant forms of the value of elementary species. He discovered and propagated (1886–1900) a series of mutations from Lamarck's Evening Primose *Oenothera lamarckiana*—a plant growing wild near Amsterdam. (Fig. 11). About 50,000 plants were cultivated, and of this number 800 were found to differ distinctly from the parent species, and to reproduce their characteristics constantly on self-fertilization. More than twelve distinct types or species were recognized by DeVries in these 800 variant forms. The following table shows clearly the result of the eight generations of a mutating strain of Lamarck's Evening Primose, and the number of forms belonging to the main new species discovered:

(1)—DeVries is Professor of Botany in the University of Amsterdam and Director of the Botanical Garden. His researches thirty years ago on osmosis in plants gave him a high reputation among plant phsiologists. He has travelled and lectured extensively in America, and has written several valuable works. (See Literature).

Generations	O. gigas	O. albida	O. oblonga	O. rubrinervis	O. Lamarckiana	O. nanella	O. lata	O. scintillans
I	......	......	......	......	9	......	......	......
II	......	......	......	......	15000	5	5	......
III	......	......	......	1	10000	3	3	......
IV	1	15	176	8	14000	60	73	1
V	......	25	135	20	8000	49	142	6
VI	......	11	29	3	1800	9	5	1
VII	......	......	9	......	3000	11	......	......
VIII	......	5	1	......	1700	21	1	......

DeVries divided these new forms into five groups:

1. *Re'rograde Varieties:*
   (1) *O. lævifolia*—smooth-leaved; strong and fertile constant.
   (2) *O. nanella*—dwarf and constant.
   (3) *O. brevistylis*—short-styled, few seeds formed, vigorous aud constant.
2. *Progressive Elementary Species:*
   (1) *O. gigas*—giant; stout; constant.
   (2) *O. rubrinervis*—red-veined and with red streaks fruit; constant.
3. *Progressive Elementary Species, but weakly*:
   (1) *O. albida*—white, narrow leaves, delicate; constant.
   (2) *O. oblonga*—narrow leaved and constant.
4. *Incomplete forms:*
   (1) *O. lata*—pistillate, low, dense.
5. *Inconstant forms:*
   (1) *O. scintillans*, dwarf and inconstant, reverting to new forms.
   (2) *O. elliptica*—narrow-leaved and inconstant.

Bateson and others suggest that some of DeVries' mutants of *Oe. lamarckiana* may be due to some sort of hybridization behavior. The new characters may be the reappearance of characters brought in by hybridization, by some process akin to Mendelian segregation.

Other suggestive reasons for believing in its possible hybrid origin are:

Davis has produced a new Oenothera by crossing *Oe. biennis* with *Oe.franciscana*, and has named it *Oe. neolamarckiana*, In its systematic characters it cannot be distinguished from *lamarckiana*, and its behavior in breeding is very similar.

Studies of *lamarckiana* seeds and pollen grains reveal a high percentage of sterility—60% of the seeds and 50% of the pollen grains—a characteristic of hybrids formed by crossing distinct species. The fact that it "produces different kinds of fertile gametes every generation, and forms *twin hybrids* in approximately equal numbers when crossed with certain wild species, as do several of the wild species in crosses with each other," shows that it has "the variability characteristic of hybrids."

Fig. 11. Lamarck's Evening Primrose ("Oenothera lamarckiana").

Examples of Mutations.—Weeping willow, nectarine, Houghton and Downing gooseberries, garden strawberry, purple leaved plum, Catawba, Concord and Clinton grapes, Wilson's Early, Lawton, Wilson Junior and Eureka blackberries, DeVries' species of Lamarck's Evening Primrose, the peloric toad-flax, cockscomb, Shirley poppy, white cyclamen, Burbank's lavender-scented dahlia, seedless navel orange, moss roses, thornless cacti, seedless banana, kohlrabi, cauliflower, red sunflower, sweet peas, Boston

Fern, White Blackbird, *Drosophila* mutations of Morgan, etc.

Fluctuating Variations vs. Mutations.—It may be instructive to summarize here the difference between fluctuating variations and mutations:—

(a The former are not transmissible; the latter are:—they breed true.

(b) The former present no new character, but the same characters differing in degree; the latter bring in a new character, or lose an old one abruptly.

(c) The former are common; the latter are rare.

(d) The former permit a series whose differences may be plotted on a frequency curve; the latter cannot be so plotted, they are discontinuous.

(c) The former fluctuate about a mean, and never produce a new permanent mean; the latter "cause a new mean to be formed, around which is grouped a new series of fluctuating variations. The real test of difference comes in breeding."

1. Discuss discontinuous variations in relation to the Biogenetic Law.
2. Are the gill-slits of embryos of mammal, bird and lizard "embryonic survivals or" "phyletic contractions?
3. What are *chimeras*? Vegetative Mutations? Give Examples.
4. Discuss lethal factors in connection with mutations.

(b)—Gates' Study of the Chromosomes of Oenotheras

Gates has recently made a careful study of the chromosomes of several species of *Oenothera*. He finds that *Oe. gigas* has 28 chromosomes as its diploid number, and as a consequence of the increased amount of chromatin the cells are larger. *Oe. lata* and *scintillans* have 15 diploid chromosomes, and as a consequence their meiotic reduction is irregular. Half of the pollen grains and half of the embryo-sacs contain 8 chromosomes, while the other half have 7 chromosomes each.

These and other facts seem to lead to the conclusion that some of the mutations are due to an irregularity in the distribution of chromosomes, a result of internal accidents.

Moreover, Gates found that the chromosomes of mutation species of *Oenothera* differ in their shape, size and structure (Gates, R.R.—*The Mutation Factor in Evolution*, 1915).

(c)—Origin of Adaptations

The question of the origin of *adaptations* to their environment has been discussed for more than fifty years, and is still unanswered. Strict followers of Darwin maintain that adaptations, indeed all specific structural differences, have slowly arisen as a result of the natural selection of useful variations as adjustments to the environment. "They explain the origin of adaptations on the basis of their usefulness" (Morgan). The weakness of their position lies in the fact that they assume the presence of useful structures without explaining how they arose. Lamarckians hold that adaptations are "the accumulation of structural responses to the conditions of the environment". While fluctuating variations do no doubt arise in this way, there is no good evidence that structural and permanent changes have arisen in the same way.

The mutationists regard adaptations as the survival of only those mutations sufficiently adapted to the environment to maintain a foothold. There is evidence that many species are poorly adapted to their surroundings, and that others are over-adapted. The former will perish where the struggle for existence becomes intense, and will be supplanted by forms that are better adapted. Again over-adapted forms can never arise by natural selection, hence such forms must have originated in some other way.

Both cases can be more readily explained by assuming that new forms are constantly arising as mutations, that those well adapted to their environment will likely survive in the struggle for existence and leave progeny, and that those not so well adapted will die but may survive for a time, where the struggle is not severe. (See page 21).

Chapter 12—JOHANNSEN'S PURE LINES (1903)

Johannsen's experiments with beans and barley, both self-fertilizing plants, seem to show that *fluctuating variations have little or no influence on the permanent improvement of a race.* He produced a number of *Pure Lines* from single plants by self-fertilization. The members of each Pure Line showed a normal variability in the weights of their seeds, which differed more or less from the mean of the variety. When a markedly divergent member of a particular line was propagated, its offspring showed regression to the mean of its particular line, but not to the mean of the variety.

Johannsen first used the term "*Pure Line*" in a pamphlet entitled *On Inheritance in Populations and in Pure Lines*, published in 1903, giving the results of a long series of experiments carried out with beans and barley. A Pure Line includes all the descendants of a single individual belonging to a strain which is reproduced exclusively by self-fertilization. It may be noted that in a Pure Line the germ plasm is never mixed, as it is derived from the same parent and there is no chance for new combinations of characters derived from different parents.

Let us examine for a moment the results obtained by Johannsen. In the case of the beans, 19 Pure Lines were grown from 19 different original plants. When all the beans

Fig. 12.—Pure lines of beans. The lower figure gives the general population, the other figures give the pure lines within the population (After Johannsen.).

were thrown together and arranged as to size a normal curve of frequency was obtained, which showed that the population belonged to a single type. The average weight was 478.9 milligrams, and the standard deviation 95.3 mgms. (Fig. 12).

When, however, the 19 separate Pure Lines were examined as to size, each had its own normal curve of frequency and its own standard deviation. Average weight ranged from 351 to 642 mgms. and the standard deviation from 64 to 109 mgms according to the following table:

Following is the result of selection in the 19 Pure Lines of Beans:

Pure Lines	Weight in centigrams of the Mother Bean						Mean Weight of the Lines
	20	30	40	50	60	70	
I	...	...	...	...	63.1	64.9	64.2
II	...	...	57.2	54.9	56.5	55.5	55.8
III	...	...	...	56.4	56.6	54.4	55.4
IV	...	...	...	54.2	53.6	56.6	54.8
V	...	...	52.8	49.2	...	50.2	51.2
VI	...	53.5	50.8	...	42.5	...	50.6
VII	45.9	...	49.5	...	48.2	...	49.2
VIII	...	49.0	49.1	47.5	...	...	48.9
IX	...	48.5	...	47.9	...	...	48.2
X	...	42.0	46.7	46.9	...	...	46.5
XI	...	45.2	45.4	46.2	...	...	54.5
XII	49.6	...	...	45.1	44.0	...	45.5
XIII	...	47.5	45.0	45.1	45.8	...	45.4
XIV	...	45.4	46.9	...	42.8	...	45.3
XV	46.9	...	...	44.6	45.0	...	45.0
XVI	...	45.9	44.1	41.0	...	...	44.6
XVII	44.0		42.4	...	...	...	42.8
XVIII	41.0	40.7	40.8	...	...	...	40.8
XIX	...	35.8	34.8	...	...	...	35.1

Next the different-sized seeds in each Pure Line, arranged in groups, were grown separately, and the progeny of each size were weighed.

Size of Mother Seeds	Average Size of Progeny
350—400 mms.	572 mgms.
450—550 mms.	549 mgms.
500—550 mms.	570 mgms.
550—600 mms.	565 mgms.
600—650 mms.	566 mgms.
650—700 mms.	555 mgms.

The results for Line II. were as follows:

Size of Mother Seeds	Average Size of Progeny
350—400 mms.	572 mgms.
450—550 mms.	549 mgms.
500—550 mms.	570 mgms.
550—600 mms.	565 mgms.
600—650 mms.	566 mgms.
650—700 mms.	555 mgms.

Johannsen also tested his Pure Lines of beans for six successive years and found that each Line kept its own individuality even when a "plus strain" was grown from the largest beans and a "minus strain" from the smallest beans.

Similar results were obtained with regard to the characters length and breadth; no effect of selection in Pure Lines could be detected. The variations were apparently caused by environmental conditions.

The conclusions[1] drawn from these and other results are that variations in the weight, length and breadth of seeds within a Pure Line are not heritable. As DeVries says, "A Pure Line is completely constant and extremely variable," that is, germ plasm is constant but fluctuating variations are common.

Discuss the value of Pure Line conception in heredity, and how does it harmonize with Galton's laws of Inheritance and Darwin's theory of Natural Selection?

Why is continuous selection within a variety necessary in some crops and not in others?

Give examples of Pure Lines in Nature.

Selection of Hooded Rats.—

Castle[1] and Phillips carried out a valuable series of selection experiments with a colony of hooded rats where the amount of pigment in the hood and the dorsal band varied. Two series of selections were made for 16 generations, one, a minus, toward a lighter type and the other, a plus, toward the darker type. In both series steady progress was made towards light in one case and dark in the other. Castle concluded from the above results that

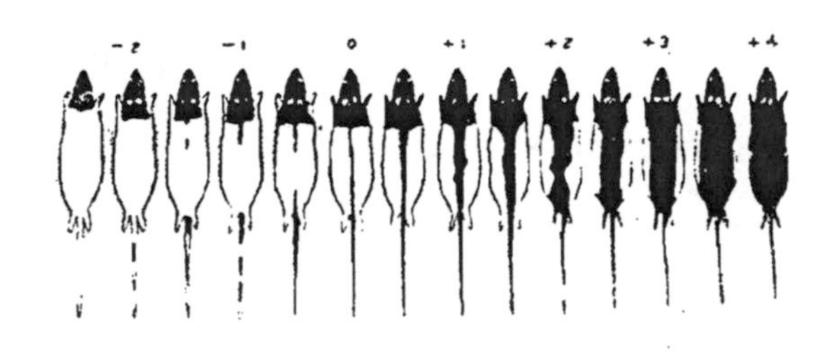

Fig. 13.—Scheme to show classes of hooded rats used by Castle. (After Castle.).

inasmuch as the hooded character is a clear example of a sharply segregating unit-character, a unit-character "is subject constantly to slight quantitative variations which are themselves to some extent hereditary." Moreover, he says "the changes affected by selection show permanency under crosses with wild races," behaving as a simple recessive unit. (Fig. 13).

(1)—See Castle's "Genetics and Eugenics" for a criticism of the Pure Line theory.

(1)—Dr. W. E. Castle, Professor of Zoology, Harvard University, has made valuable contributions to our knowledge of invertebrates, and to the more intricate aspects of heredity and gentics.

Selection of Paramoecia.

Jennings[1] isolated from a population of *Paramœcia* eight races of Pure Lines, and his results confirm those of Johannsen.

Selection of Leptinotarsa.

Tower's[2] results in breeding experiments with races of *Leptinotarsa 10-lineata* are also in agreement with Johannsen's. Dark and light colored variations appeared in the same Pure Line. When dark males and females were mated the progeny were not dark, but fluctuated about the average of the Pure Line, even after twelve generations of such mating.

Pure Lines and Mendelism.—It is now known (see pages 81-88) that the F_1 generation of a Mendelian crossing all progeny are hybrid, in the F_2 generation half are hybrid and half are pure regarding any particular character. In the F_3 generation only one fourth are hybrid in the case of self-fertilizing plants, and in the F_4 generation only one-eight are hybrid. In other words a self-fertilizing population soon becomes composed of a number of pure lines, a result which has been verified by experimenters.

Chapter 13—TYPES OF BISEXUAL INHERITANCE

With regard to Bisexual Inheritance three types may be distinguished:

1. *Blended Inheritance,*[3] where there appears to be a fusion of two characters when brought together in transmission. As examples we have the familiar cases of the blending of colors in mulattos, in horses, in cattle, in stature of man, and perhaps in prepotency.

2. *Mosaic or Particulate Inheritance,* where the characters are separately expressed in different parts of the same organ or system. Common examples are found in piebald horses, cattle, and dogs.

3. *Alternate or Exclusive Inheritance,* where the character of one parent is expressed apparently to the complete exclusion of that of the other. Examples are found in

(1)—Dr. H. S. Jennings (1868-), Professor of Zoology, University of Pennsylvania, was educated at the Universities of Michigan, Harvard, and Jena. He is recognized as the foremost investigator of the behavior of organisms.

(2)—Dr. W. L. Tower (1872-), of the University of Chicago, was educated at Harvard and Chicago. His studies on the evolution of chrysomelid beetles and the development of coloration in insects are masterpieces.

(3)—Galton drew a distinction between Blended and Particulate Inheritance, but regarded all inheritance as "largely, if not wholly, particulate."

large numbers in the study of Mendelism, and recent studies reveal the fact that many cases of Blended and Mosaic Inheritances also act according to Mendelian laws. Examples of these are given in the Chapter on Mendelism.

Chapter 14.—HYBRIDS AND HYBRIDIZATION

Hybrids in the modern sense are the product of crosses between individuals belonging to different races or elementary species. Such races or species may belong to different Linnæan genera, species or varieties. But it is not possible to produce hybrids from any two races of plants or animals taken at random, and systematic relationship is not always a reliable guide to capacity for hybridization. This capacity is usually restricted to nearly related plants and animals. As a rule, the closer the alliance is, the greater is the capacity for crossing. Experiments go to show that crosses between species are not rare, as for example, brown and polar bear, horse and zebra, horse and ass, cow and bison, duck and goose, canary and finch, pheasant and fowl, thrush and blackbird, raspberry and blackberry, wheat and rye, species of strawberry, etc.

The development of our knowledge of hybridization is largely due to Kœlreuter (1733-1806), Knight (1758–1843), Gærtner and Darwin. Later additions to our knowledge were contributed by Naudin, Focke, Vilmorin, Mendel, DeVries and many others. Fairchild, an Englishman, was the first gardener to raise a hybrid, in 1817, by pollinating the stigma of *Dianthus caryophyllus* with pollen from *D. barbatus.* Kœlreuter was the first botanist to carry on experiments on hybridization on a large scale for years. His first hybrid, *Nicotiana rustica x N. paniculata* was produced in 1761. He made an important discovery in plant breeding, that the product of reciprocal crosses are identical. Moreover, he established on a firm basis the theory of sexuality of plants.

Thomas Andrew Knight produced several commercial varieties by crossing and has been called the father of modern plant breeding. He enunciated two important principles, viz. (1) food supply is the main cause of variation, and (2) crosses are often more vigorous than the parents.

Carl Friedrich Gærtner, son of Joseph Gærtner a distinguished botanist, carried out a very large number of experiments (nearly 10,000) in crossing, and the results were published in 1838 and 1849. He divided his hybrids into *intermediate types, commingled types*, and *decided types*.

He noted, moreover, various forms in hybrids of the second generation, but he did not see segregation. He noted also the increased vigor of growth of many hybrids of species. Gærtner's crossings were made upon species as units and not upon the basis of unit-characters.

Of the many problems of hybrids and hybridization that presented themselves to Darwin, some have been solved by the aid of Mendel's Laws, but some still await solution. Among these may be mentioned: the cause of sterility; why the hybrids of reciprocal crosses sometimes differ in fertility; why some forms profit by crossing while others do not; why a change of environment affects the sterility of self-sterile species.

It is worth while noting some of Darwin's observations which come very near to those expressed by Mendel. He says;"There are certain hybrids which instead of having, as is usual, an intermediate character between the two parents, always closely resemble one of them" *(Origin of Species)*. In crossing the normal snapdragon with the abnormal he found all the hybrids normal, but when these hybrids were selfed he obtained normal and abnormal forms in the ratio of 88 to 37—a close approximation to Mendelian expectations.

He recognized latent characters, and explained "reversion" as due to the sudden reappearance of some latent ancestral character. He recognized segregation in the second generation, for he says: "The grandchildren and succeeding generations (of crosses) continually revert in a greater or lesser degree to one of both of their progenitors (*Variation of Plants and Animals under Domestication*).

Darwin maintained that the benefits from cross-fertilization depend on the fact that plants taking part in the cross have become differentiated by exposure to different conditions.

He also drew attention to the fact of immediate improvement due to a cross. He says: "After plants have been propagated by self-fertilization for several generations, a single cross with a fresh stock restores their pristine vigor" (*Effects of Cross and Self Fertilization in the Vegetable Kingdom*).

Darwin's Theory as to the combination of characters in hybridization is stated as follows:—" *When two hybrids pair, the combination of pure gemmules derived from the one hybrid with the pure gemmules of the same parts derived from the other would necessarily lead to complete reversion of character*, and it is perhaps not too bold a supposition that unmodified and undeteriorated gemmules of the same nature would be especially apt to combine.

"Pure gemmules in combination with hybridized gemmules would lead to partial reversion, and lastly, *hybridized gemmules derived from both parent-hybrids would simply reproduce the original hybrid form.* All these cases and degrees of reversion incessantly occur" (*Variation of Plants and Animals under Domestication*).

This statement comes very close to the modern theory of *heterozygosis.*

To Darwin we owe the phrase: "Nature abhors perpetual self-fertilization," which, of course, does not hold true in the case of many vigorous plants, such as tobacco, wheat, and barley. East and Jones state that Darwin's phrase should be changed to read: "Nature discovered a great advantage in an occasional cross-fertilization." It is well known that many plants have special adaptations in their flowers, whereby self-fertilization is prevented; that highly colored flowers are usually cross-fertilized by insects; that the more inconspicuous flowers are cross-pollinatd by wind, etc., etc. Darwin proved by numerous experiments that the products of crosses were usually more vigorous than the parents of the hybrid (*Cross and Self Fertilization in the Vegetable Kingdom*).

In one case he grew five pots of cross-fertilized seeds of pansy and an equal number of self-fertilized seeds with the following results:

Average height of the 14 cross fertilized plants =5.58 inches.

Average height of the 14 self-fertilized plants = 2.37 inches.

Next year the crossed plants produced 167 capsules; the selfed plants produced 17 capsules.

(Plants turned out of pots and planted in soil).

In the following year the area covered by the cross-fertilized plants was about nine times as large as that covered by the self-fertilized plants. During the ensuing winter, which was severe, all the cross-fertilized plants survived and grew vigorously.

Again the seeds from cross-fertilized flowers on plants that came from self-fertilized flowers were planted. The results were:

Average height of six plants from first set = 10.31 inches.

Average height of six plants from second set = 12.56 inches.

Moreover, there is a great range in the productivity of hybrids— from absolute sterility[1] to complete fertility. In this regard also hybrids of closely related parents are, as a rule, more fertile than those from widely different parents.

Species-hybrids have been divided into three classes:

1. Where the hybrid has vigor and fertility equal to or greater than the parents, e. g. *Nicotiana alata x N. langsdorffi*.

2. Where the vigor is equal to or greater than that of the parents but the fertility is much reduced, e. g. *Raphanus sativus* x *Brassica oleracea; Bison americanus* x *Bos taurus*.

3. Where both size and vigor have become much reduced and there is complete sterility, e. g. *Nicotiana tabacum* x *N. paniculata; N. rustica* x *N. alata*.

Hybrids between nearly related but *distinct* species are frequently observed to have characters *intermediate* between the unlike characters of the parents; they are *vigorous and stable*. but more or less sterile. On the other hand, hybrids between doubtfully distinct species, or between a species and a variety, do not have *intermediate* characters but, as a rule, act in a Mendelian manner.

The large number of stable species in the *Onagraceæ, Solanaceæ, Rosaceæ, Compositæ, etc.* is accounted for by some investigators by the formation of natural hybrids of distinct species.

Jeffrey, making pollen sterility a criterion of hybridization of species, is of the opinion that DeVries' *Oe-lamarckiana* is a hybrid and not a true species.

East and Shull discovered an interesting point regarding the influence of continued self-fertilization and crossing. They found that although corn will lose vigor for several generations when self-fertilized, the loss does not continue at the same rate—decreasing in successive generations until a condition of constancy is reached. In this condition cross-fertilization with a plant of the same strain does not increase the vigor of the progeny.

(1)—Infertile hybrids may be of great commercial value if they can be reproduced by cuttings, grafts, tubers, etc.

If, however, cross-fertilization occurs with a different strain, an immediate increase of vigor follows. Shull's results were as follows:

Average yield per acre of two pure strains. . 29 bushels
Yield of first cross.................... 68 bushels.
Yield of original strain (cross-fertilized for 5 years) 61½ bushels.

Here the problem is to determine the real reason for the increased vigor in the first cross of two separate strains.

Additional examples of the increased vigor of hybrids over the parent forms are Burbank's hybrid walnut (California black walnut x English walnut) which grows larger and much faster than either parent, the hybrid dewberry (Western dewberry x Siberian raspberry) which ripens earlier and is larger and more productive than either parent.

As examples of the increased size of hybrids among our domestic animals we may note that crosses between Cheviots and Leicesters are larger than either parent, and such crossings are much practised when lambs and sheep are reared for the market (See also Chapter 17 on In-breeding); that the mule is extensively bred in the United States on account of its hardiness in extreme climatic conditions, its longevity, its relative freedom from disease and injury, and its ability to live on coarse food; and that the first crosses between Poland-China and Chester-white swine, and between Shorthorn and Aberdeen-Angus cattle are popular as feeders.

The work of Webber and Swingle with citrus fruits is also worthy of mention. By crossing the ordinary orange with the hardy trifoliate orange, a useless variety, they secured among the hybrids several that combined quality with frost resistance. These hybrids are often called "citranges." The Rusk, the Willets and the Morton are three of the new fruits. As a result oranges can be grown 400 miles north of their present range, and able to endure a temperature of 8° F.

Another hybrid the "tangelo" resulted fron crossing the tangerine with grape fruit.

The investigations by D. F. Jones showed that grains of hybrids weighed more than those of the self-fertilized forms.

Shull has attempted to explain hybrid vigor or heterosis by the theory of *heterozygosis*, *i.e.* that hybrid vigor is in proportion to the number of factors in which the parents differ. Obviously in crossing two strains with different factors such as:

AA, BB, CC, DD x aa, bb, cc, dd.

the hybrid will be Aa Bb Cc Dd, which is completely heterozygous. If this is "selfed" homozygous conditions are provided with regard to one or more factors with attendant loss of vigor. After a time if "selfing" be continued, a perfectly homozygous condition is reached.

Again, the results suggest that the vigor increases with the number of 'desirable' factors present in the hybrid. Jones' theory of *dominance of linked characters* appears quite plausible as an explanation (See Genetics 2,197).

Success in hybridization demands technical skill and accurate knowledge as to the pollination habits of plants. The flower buds are opened carefully and the anthers removed, before they mature, with a pair of scissors or forceps. They are then enclosed in a loose paper bag to prevent the entrance of foreign pollen. When the stigmas are mature ripe pollen is transferred from the anthers of the male parent by means of a brush or by direct contact of anther and stigma, and the paper bag is again fastened over the flowers until fertilization is effected.

Some plants are normally self-fertilized, while others are normally cross-fertilized. There are many plants, however, that occupy an intermediate position—some self-fertilizing plants being sometimes cross-fertile, and some cross-fertilizing plants sometimes self-fertile. Following is a list of the common economic plants arranged according to pollination habits:

1. *Self-fertilized plants.*—

(a) Normally self-fertilized:—wheat, barley, oats, rice, beans, peas, and most legumes.

(b) Sometimes cross-fertilized:—Tobacco, tomato, flax, cotton.

2. *Cross-fertilized plants.*—

(a) Also self-fertilized:—rye, sugar beet, some pears, corn and cucurbits.

(b) Self-sterile:—white and red clovers, alfalfa, sunflower, most apples, some pears, plum, cherry gooseberry, currant.

(c) Diœcious:—asparagus, hemp, hops.

Success requires, moreover, the use of large numbers of plants so that an individual with the desired combination of qualities may be secured. Burbank has used this method with advantage producing, among others, the seedless apple, the large white blackberry, the spineless cactus, the Shasta Daisy (a triple hybrid of American, English and Japanese daisies) the rapid growing Royal walnut, (a hybrid of *Juglans californica* x *J, nigra*), and the dewberry-raspberry, the plumcot (a hybrid of plum and apricot.)

East is of the opinion that this phenomenon of increased vigor of hybrids may account for the frequency of cross-fertilized species and the rarity of self-fertilized species, since it can be shown that there is no evil effect due to inbreeding *per se.*

Exercise.—Give cases of increased hybrid vigor, other than those cited above.

Uses of Hybrids:

Summarizing we may say that hybrids are produced:

1. to obtain a combination of desirable characters not present in other plants, e.g. many of Burbank's productions.
2. to get rid of undesirable characters such as susceptibility to disease, drought and cold, e. g. work of Biffen with wheat and Webber with critus fruits; and
3. to secure greater vigor and more rapid growth, e.g. work of Burbank with walnut trees, and that of East and Shull with corn.

Chapter 15—THE EXPERIMENTAL METHOD OF INVESTIGATING HEREDITY

The second method of studying heredity is by direct experimentation after the method of Mendel and his followers. (See p. 46)

(a)—Mendelism

The laws of Inheritance in hybridism were first determined by Abbe Gregor Mendel[1] and published in 1865 in the Proceedings of the Natural History Society of Brünn[2]

(1)—Mendel was born at Heinzendorf in Austrian-Silesia in 1822 and was educated at the Gymnasia of Troppau and Olmutz. He entered the Augustinian Monastery at Brunn where he became a noted teacher. He studied mathematics, physics and natural sciences for two years (1851-1853) at the University of Vienna, and in 1868 became Abbot of the Monastery. He corresponded with the Botanist Nageli who was the only great naturalist acquainted with his experiments on peas. He died in 1884.

(2)—The French botanist Naudin published in 1862 the results of researches in plant hybridization which harmonized with those of Mendel, but he failed to grasp the idea that the segregation applies to "single" characteristics rather than to all the characteristics of a species at once! No doubt Mendel was acquainted with his researches as well as those of his predecessors, Kolreuter, Knight and Goss (See Lock).

This discovery, however, remained unnoticed until 1900, when DeVries of Holland, Correns of Germany, and Tschermak of Austria rediscovered the laws simultaneously. In this year too, Mendel's publication came to light.

These laws have been stated as follows:

1. "If two contrasted characters, which have previously bred true, are crossed, only one—the dominant character—appears in the hybrid (*Law of Dominance*)."

2. "In succeeding generations self-fertilized plants grown from seeds from this cross reproduce both characters in the proportion of three of the dominant character to one

Fig. 14.—GREGOR MENDEL about the year 1862.

of the recessive character. Furthermore, the recessive character continues for ever after to breed true, while those plants bearing the dominant character are one-third pure dominants, which ever after breed true to the dominant character, and two-thirds hybrid dominants which contain the recessive character in a hidden condition (*Law of Segregation*." (Fig. 14).

Explanation of the Laws. The results of his experiments (see below) led Mendel to the conception of pairs of unit characters, or allelomorphs, of which either can be carried to any gamete or sex cell to the exclusion of the other.

As the possession of either character is a matter of chance, on the average 50 per cent. of the germ cells in the hybrid will bear the dominant character and 50 per cent. will bear the recessive character. If we could pick out at random any 100 pollen or male cells to fertilize any 100 egg or female cells we will see that there are equal chances for four results:

1. A male cell with a *dominant* character may meet a female cell with a dominant character.
2. A male cell with a *dominant* character may meet a female cell with a *recessive* character.
3. A male cell with a *recessive* character may meet a female cell with a *dominant* character.
4. A male cell with a *recessive* character may meet a female cell with a *recessive* character.

In an abbreviated form the mating may be represented as follows:—

Male Cells		Female Cells
D	....	D
D	....	D
R	....	R
R	....	R

or 1 DD—2 DR—1 RR

The net result will, therefore, be 25 per cent. pure dominant, 25 per cent. pure recessive, and 50 per cent. impure dominant.

Mendel's Experiments.—Mendel used pure strains of the common garden pea (*Pisum sativum*) in his memorable experiments which led to the discovery of the laws of inheritance. Twenty-two varieties or sub-species were selected and used during the eight years of experimentation. Seven different characters were chosen and investigated separately:

1. *Form of seed*—round or wrinkled[1]
2. *Color of endosperm*—yellow or green

} Cotyledon or first generation characters

3. *Color of seed coat*—white or grey
4. *Form of pods*—soft or hard
5. *Color of unripe pods*—green or yellow
6. *Position of flowers*—axial or terminal
7. *Length of stem.*—tall or dwarf.

} Parental characters

(1)—Round and wrinkled are termed contrasted characters; likewise yellow and green; white and grey, etc.

In the first generation only one of each pair of contrasted or mated characters appeared. Such was termed the dominant character. In the next generation, the hybrids being self-fertilized, there occurred a splitting or segregation into the original characters according to a definite ratio. For example, when a yellow-seeded variety was crossed with a green-seeded variety, the hybrids were all yellow-seeded—yellow being dominant and green recessive. When the yellow-seeded hybrids were self-fertilized, the progeny consisted of 75 per cent. yellow seeds, and 25 per cent. green seeds. The green-seeded plants, however, when self-fertilized, produced green seeds entirely, hence were pure; but when the yellow-seeded plants, were self-fertilized some of the progeny produced yellow and green seeds, while others produced yellow seeds only.

The results may be shown diagrammatically as follows:

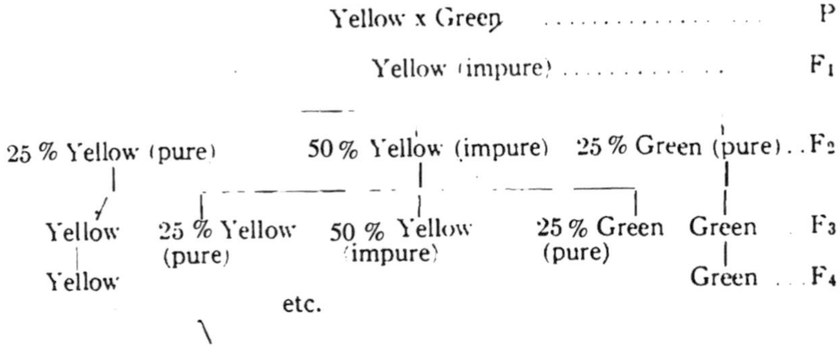

Yellow x Green . . . . P

Yellow (impure) . . . . F_1

25 % Yellow (pure) | 50 % Yellow (impure) | 25 % Green (pure) . . F_2

Yellow | 25 % Yellow (pure) | 50 % Yellow (impure) | 25 % Green (pure) | Green . . F_3

Yellow | | | | Green . . F_4

etc.

The gametic matings in the "selfing" for the F_2 generation may be represented as follows when it is borne in mind that in fertilization there will be two kinds of gametes and these in equal numbers:

		Y	G	
Female gametes in equal numbers	Y	YY	YG	= male gametes in equal numbers
	G	YG	GG	= YY+2YG+GG

Where YY is a pure yellow zygote,[1] YG is an impure yellow, and GG a pure green, or 25 % pure yellow, 50 % impure yellow, and 25 % pure green.

(1)—The pure yellow zygote (YY) formed by the union of two like gametes is termed a "homozygote" and the impure yellow zygote (YG) formed by the union of two unlike gametes a "heterozygote." These two zygotes are "phenotypes," but genetically they belong to different "genotypes."

Summary of Mendel's Experiments with Peas:

No.	Character contrast	No. in F2	Dominants	Recessives	Ratio per 4
1	Form of Seed	7,324	5,474	1,850	2.99:1.01
2	Color of Cotyledons	8,023	6,022	2,001	3.00:1.0
3	Color of Seed-coats	929	705	225	3.04:0.96
4	Form of Pod	1,181	882	229	2.99:1.01
5	Color of Pod	580	428	152	2.95:1.05
6	Position of Flowers	858	651	207	3.03:0.97
7	Length of Stem	1,064	787	277	2.92:1.08
	Totals.	19,959	14,949	5,010	2.996:1.004

The foregoing results were obtained when one pair of characters was used, but Mendel found the law to hold true when two or more sets were considered at the same time.

FIG. 15.
MENDELIAN PROPORTIONS IN MAIZE.
Cobs born by heterozygote plants pollinated with the recessive, showing equality of smooth and wrinkled and of colored and white grains.

The resu'ts, however, are more complicated, but they show that each set of characters acts independently of the others in the inheritance. (See section on Dihybrids, page 87).

To Identify a Heterozygote.—In practical breeding it is often necessary to know definitely whether the organism is homozygous or heterozygous as to a certain character. The easiest way of identification is to cross with a recessive. If the organism is a homozygote all the progeny will show dominance, if a heterozygote only fifty per cent. will show dominance, according to the following diagram:—

(1)

		D	D	= male gametes
female gametes	R	RD	RD	= 4RD, or all showing dominance.
	R	RD	RD	

(2)

		D	R	= male gametes
female gametes	R	RD	RR	= 2RD + 2RR, or 50% dominant. (Fig. 15)
	R	RD	RR	

How Unit Characters may be Determined.—The only method of determining unit characters is by crossing. Mendelian investigators have revealed the presence of pairs of contrasting characters when varieties are crossed both in the F_1 and F_2 generations.

(1) The hybrids of crosses (F_1) frequently show one of the pairs of characters when dominance exists.

(2) Segregation or splitting occurs in the F_2 generation.

Simple Mendelian Segregation

Before beginning the study of segregation in di-hybrids and tri-hybrids it is advisable to become familiar with the results of simpler crosses: (1) a *homozygote*, such as a "pure yellow" pea, with a *heterozygote*, such as an impure yellow.

All the male gametes will contain the factor for yellow, and of the female gametes half will contain the factor for yellow and half the factor for green. The possible matings may be represented as follows.—

		Y	Y	= Male Gametes.
Female Gametes	Y	YY	YY	
	G	GY	GY	= 2YY + 2GY; Phenotypically all yellow, but genotypically half are pure yellow and half impure yellow.

(2) a *heterozygote* yellow with a *homozygote* green. Of the male gametes half will contain the factor for yellow, the other half for green; all the female gametes will contain the factor for green. The possible matings may be represented as follows:

	Y	G	= Male Gametes.
Female Gametes G	YG	GG	= 2 YG + 2 GG
Female Gametes G	YG	GG	

Phenotypically half are yellow and half are green, but genotypically half are impure yellow and half are pure green.

Two Pairs of Characters in Dihybrids

When two pairs of characters hybridize, such as wrinkled-green peas with smooth-yellow, the Mendelian law still holds good. On the assumption that a gamete can contain only one of each pair of factors, one-fourth of the male gametes will contain the factors for *smoothness and greenness* (SG), one-fourth the factors for *smoothness and yellowness* (SY), one-fourth the factors for *wrinkledness and greennes* (WG) and one-fourth for *wrinkledness* and *yellowness* (WY).[1] Similarly for the female gametes. When these gametes unite, the possible combinations may be represented as follows:—

Female Gametes	SG	SY	WG	WY	Male Gametes
SG	SSGG	SWGG	GWGG	SWYS	
SY	SSYG	SSYY	SWYG	SWYY	
WG	SWGG	SWYG	WWGG	WWYG	
WY	SWYG	SWYY	WWYG	WWYY	

or SSGG + 2SSYG + 2SWGG + 4SWYG + SSYY + 2SWYY + 2WWYG + WWGG + WWYY—nine different combinations or genotypes. However, as S is dominant to W, and Y to G, there will be 9 smooth-yellow, 3 smooth-green, 3 wrinkled-yellow and 1 wrinkled-green phenotypes, four in all. The ratio is, therefore, 9:3:3:1. (Fig. 16).

(1)—In the calculation of Mendelian results, Castle gives this important bit of advice: "It is all essential to determine first the kinds of gametes each parent is expected to produce. The subsequent calculation is easy and certain."

Several investigators have discovered that the ratio 9:3:3:1 does not always occur in di-hybrids. This deviation is now explained by Morgan as due to *gametic coupling* or crossing-over of the chromosomes in the synapsis stage of nuclear formation. (See page 123).

Problem.—Determine the F_1 and F_2 generations when black hornless cattle are crossed with red horned cattle, when black is dominant to red, and hornless to horned.

Problem 2.—Determine the F_1 and F_2 generations when smooth white guinea pigs are crossed with rough black guinea pigs, when rough and black are dominant.

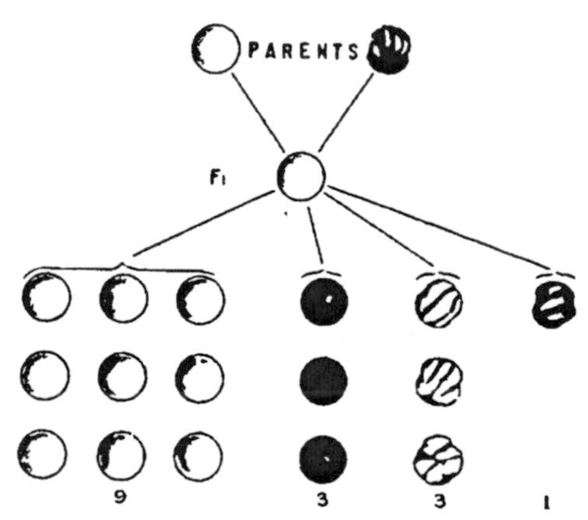

Fig. 16 -- Cross between yellow-round and green-wrinkled peas, giving the 9:3:3:1 ratio in F2. (Morgan)

Three Pairs of Characters in Hybridism—Trihybrids

Mendel showed that when three pairs of characters are involved the law still holds goods. In the case of peas, when *yellow-smooth-tall* peas were crossed with *green-wrinkled-dwarf* peas the hybrids were all *yellow-smooth-tall.*

On the assumption that a gamete can contain only one of each pair of characters the probabilities are that there will be an equal number of eight kinds of male gametes in the hybrid: YST, YSD, YWT, YWD, GST, GSD, GWT, a GWD. Similarly for the female gametes. Accordingly the zygotes for the F_2 generation will be constituted as follows:

ZYGOTES OF THE F_2 GENERATION IN TRI-HYBRIDS

Male Gametes		YST	YSD	YWT	YWD	GST	GSD	GWT	GWD
Female Gametes	YST	$Y^2S^2T^2$	Y^2S^2TD	Y^2SWT	Y^2SWTD	YGS^2T^2	YGS^2TD	$YGSWT^2$	YGSWTD
	YSD	Y^2S^2TD	$Y^2S^2D^2$	Y^2SWTD	Y^2SWD^2	YGS^2TD	YGS^2D^2	YGSWTD	$YGSWD^2$
	YWT	Y^2SWT^2	Y^2SWTD	$Y^2W^2T^2$	Y^2W^2TD	$YGSWT^2$	YGSWTD	YGW^2T^2	YGW^2TD
	YWD	T^2SWYD	Y^2SWD^2	Y^2W^2TD	$Y^2W^2D^2$	YGSWTD	$YGSWD^2$	YGW^2TD	YGW^2D^2
	GST	YGS^2T^2	YGS^2TD	$YGSWT^2$	YGSWTD	$YG^2S^2T^2$	G^2S^2TD	G^2STW^2	G^2SWTD
	GSD	YGS^2TD	YGS^2D^2	Y^2SWTD	$YGSWD^2$	G^2S^2TD	$G^2S^2D^2$	G^2SWTD	G^2SWD^2
	GWT	$YGSWT^2$	YGSWTD	YGW^2T^2	YGW^2TD	G^2SWT^2	G^2SWTD	$G^2W^2T^2$	G^2W^2TD
	GWD	YGSWTD	YGSWDD	YGWWTD	YGWWDD	GGSWTD	GGSWDD	GGWWTD	GGWWDD

Y = factor for yellow; S = factor for smooth; T = Factor for tall.
G = Factor for green; W = Factor for wrinkled; D = Factor for dwarf.
A zygote with the factor constitution of (say) YGSWTD is yellow, smooth, tall; since these factors are dominant to green, wrinkled and dwarf respectively.

Of these 64 zygotes one is pure yellow—smooth—tall; one pure yellow—smooth—dwarf; one pure yellow—wrinked—dwarf; one pure green—smooth—tall; one pure green—smooth—dwarf; one pure green—wrinkled—tall; and one pure green—wrinkled—dwarf. The remainder are impure forms.

Phenotypically however, there are 27 yellow—smooth—tall, 9 yellow—smooth—dwarf, 9 yellow—wrinkled—tall, 9 green—smooth—tall, 3 yellow—wrinkled—dwarf, 3 green—wrinkled—tall, 3 green—smooth —dwarf, and one green—wrinkled—dwarf,—the ratio being 27 : 9 : 9 : 9 : 3 : 3 : 3 : 1. On examination of the zygotes it will be found that there are 27 different genotypes, and 8 phenotypes.

Explain how Biffen was able to breed a wheat that was *hard, heavy-yielding,* and *immune to rust* from a variety that was *soft, heavy-yielding,* and *susceptible to rust,* when hardness and susceptibility to rust are dominant characters.

Explain the following formulæ for segregation given by Morgan:

3		1		One pair of factors.
9 : 3		3 :1		Two pairs of factors.
27:9	9:3	9:3	3:1	Three pairs of factors.

It is important to observe that the first two pairs of characters in Mendel's list (page 83) pertain to the cotyledons which form the first or seed leaves of the next generation, and the following five pairs of characters pertain to the seed coat, pod, flower or stem, which belong to the parental generation. In actual breeding operations, therefore, when a pure yellow pea is crossed with a green pea the yellow color of the cotyledons of the hybrid will be obtained the same year. But when a pure tall pea is crossed with a dwarf pea, the tall hybrid will not be obtained until the following year after the seed is sown.

Similarly when a tall yellow pea is crossed with a green dwarf pea, the characters of the yellow tall hybrid will not find expression at the same time.[1]

Mendel chose the garden pea because it is easy to cultivate, has a short period of growth, and has constant differentiating characters. With true scientific insight he paid attention to the following problems :

1. " To determine the number of different forms under which the offspring of hybrids appear.

(1) (See also Vol. XI. No. 4, "Journal of Agricultural Research" on the Researches on the 35 Factors of the Genus Pisum).

2. To arrange these forms with certainty according to their generations.
3. To ascertain accurately their statistical relations." (From Mendel's original Paper.)

(b)—Dominant and Recessive Characters

Experiments go to show that the Mendelian laws of inheritance apply to a great many characters, but *dominance* is not considered now an essential feature.

Castle gives the following list of dominant and re-recessive characters:—

DOMINANT	*RECESSIVE*
CATTLE:	
Black	Yellow
Polled	Horned
Dexter form (short legs)	Kerry form (legs normal)
Dominance uncertain or variable—	
White	Colored
Uniformly colored	Spotted with white
Uniformly black	Black spotted with yellow
HORSES:	
Bay	Not bay (i.e.black or chestnut).
Black	Chestnut.
Gray.	Not gray (any color but gray).
Trotting.	Pacing.
Dominance uncertain or wanting—	
Uniformly colored.	Spotted with white
SWINE[1]:	
Wild color.	Not wild color (black or red).
Black.	Red.
Self white.	Colored.
Mule-footed (syndactyly)	Normal foot.
Dominance uncertain or wanting—	
Uniformly colored	Spotted with whi e.
SHEEP:	
White fleece.	Black fleece.

(1)—See also Journal of Heredity, Vol. VIII, No. 8.

DOMINANT	*RECESSIVE*
DOGS:	
Gray	Black.
Self Color.	Bi-color (black and tan, brown and tan, red and tan).
Black.	Yellow or(red).*
Black.	Brown (liver).
Dominance uncertain or wanting—	
Colored all over.	Spotted with white
Black or Brown.	Black or brown spotred with yellow.
Stumpy tail.	Normal tail.

(*In Dachshunds red is not uniformly recessive; it apparently may be dominant).

DOMINANT	*RECESSIVE*
CATS:	
Tabby.	Not tabby (black or blue).
Black	Blue.
Short hair.	Long hair (angora)
Dominance imperfect or uncertain—	
Colored all over.	Spotted with white.
White (eyes only colored).	Colored all over.
Yellow.	Not yellow (tabby or black)
Tailless.	Long-tailed.
Polydactyly.	Toes normal.
RODENTS:	
Colored.	Albino.
Black or Brown.	Yellow.
Gray (agouti).	Black or brown (non-agouti).
Black or black agouti.	Brown or Brown agouti.
Self colored.	Spotted with white.
Self black or brown or agouti.	Black, brown or agouti spotted with yellow.
Dark eyes and coat.	Pink eyes and coat, pale where not yellow.
All pigments dark.	All pigments pale.
Short-haired like wild cavies.	Hair long and silky.
Coat rosetted.	Coat smooth.
Unbanded shell in wood-snail.	Banded shell.
Yellow cocoon in silkworm.	White cocoon.

DOMINANT	*RECESSIVE*
FOWLS:	
Jungle-fowl color pattern.	Self black or yellow.
White (of white Leghorns)	Colored.
Colored.	White (of "silkies" and some other white breeds).
Barred.	Not barred.
Black plumage.	Yellow plumage (hetero-zgote often like jungle fowl).
Black skin.	Normal skin.
Crest.	No Crest.
Frizzled.	Not frizzled.
With extra toe.	Without extra toe.
Walnut comb.	Pea, rose or single comb.
Pea comb.	Single comb.
Rose comb.	Single comb.
PLANTS:	
1. Colors of flowers. etc. (Example, unit-characters of the sweet pea flower).	
Colored.	White.
Colored.	Slightly colored (picotee).
Purple.	Red.
Bi-color.	Self.
Oval pollen grains.	Round pollen grains.
2. Forms of flowers	
Normal.	Peloric.
Single.	Double.
Long style of Oenothera	Short style.
Entire petals of Chelidonium.	Laciniate petals.
3. Colors of Leaves and Stem.	
Variegated with yellow.	Normal green (dominance imperfect.)
Containing much red.	With little red (Oenothera, Coleus, maize).
4. Colors of Fruits and Seeds. (Example, maize).	
Yellow endosperm.	White endosperm.
Aleurone black.	Aleurone red or uncolored.
Aleurone red.	Aleurone uncolored.
Endosperm starchy.	Endosperm sugary.
Endosperm starchy.	Endosperm waxy.
Seed-coat red.	Seed-coat colorless.
Seed-coat variegated.	Seed-coat not variegated.

DOMINANT	*RECESSIVE*
5. Forms of Leaves.	
Serrate.	Entire (Urtica).
Normal.	Laciniate (Chelidonium).
Palmate.	Pinnatifid or fern-leaf (Primula).
Hairy.	Glabrous (dominance often imperfect).
WHEAT:	
Absence of awn.	Presence of awn.
Rough and red chaff.	Smooth and white chaff.
Keeled glumes.	Rounded glumes.
Flinty endosperm.	Floury endosperm.
Susceptibility to rust	Immunity to rust.
BARLEY:	
Two-rowed ears.	Six-rowed ears.
Beardlessness.	Bearded.
COTTON:	
Colored lint.	White lint.
TOMATOES (after Price and Drinkard)	
Fruit Shape:	
Spherical or round.	Pyriform.
Two-celled.	Many-celled.
Roundish-conic.	Roundish compressed.
Fruit Color:	
Red Fruit	Pink Fruit.
Red Fruit.	Yellow Fruit.
Pink Fruit.	Yellow Fruit.
Yellow Fruit Skin.	Transparent Fruit Skin.
Fruit Surface:	
Smooth.	Pubescent.
Foliage:	
Normal or cut leaf.	Potato Leaf.
Pimpinellifolium Leaf.	Normal Leaf.
Green Leaf.	Yellow Leaf.
Normal or Smooth Leaf Surface.	Rugous Leaf.
Stature:	
Standard Stature.	Dwarf Stature.

Most mutants are recessives when mated with normal allelomorphs but numerous examples can be cited which show dominance.

Meaning of Dominant and Recessive

Confusion has arisen in some quarters as to the meaning of the terms *dominant* and *recessive*. For example, it has been suggested that "a dominant character is the result of a long process of selective breeding." This idea of course is groundless.

Again, it has been stated more frequently that the older of the two members of a pair of characters is dominant. This theory holds good in some cases but it breaks down in others.

Technical Terms.—In the great extension of our knowledge of heredity during the last few years certain terms have been introduced to express more clearly and accurately the new and complex situations that have arisen. Instead of the term "character," "determiner" is used as being that element or condition within the gamete which determines the development of a given somatic character. Synonymous terms are *factor* and *gene*. (See also page 109).

"Dominant" and "recessive" characters are now frequently explained by the "*Presence* and *Absence Theory,*" which assumes that a determiner for any character either is, or is not, present,

If both of the pairing gametes contain the same determiner, the *zygote* or offspring will be *homozygous* or will have a double, or *duplex*, dose of the character. If only one of the pairing gametes contains the determiner, the zygote will be *heterozygous* or will have a single, or *simplex*, dose of the character. If absent from both gametes the zygote is said to be *nulliplex*.

Organisms that have identical determiners in the germ-plasm belong to the same *genotype*, while those that appear alike without reference to germ plasm belong to the same *phenotype*. (See also Glossary, page 177).

Exercise.—Determine the probable nature of the progeny when

(1) Duplex x duplex =
(2) Duplex x Simplex =
(3) Simplex x Simplex =
(4) Duplex x Nulliplex =
(5) Simplex x Nulliplex =
(6) Nulliplex x Nulliplex =

The F_2 Generation

The results of simple Mendelian segregation in the F_2 generation may be summarized as follows:

The F_2 Generation or the Progeny of Hybrid x Hybrid:

No. of Pairs of Parental Characters.	No. of Kinds of Gametes	No. of Genotypes	No. of Phenotypes	Ratio of number of Phenotypes.
1	2	3	2	3: 1
2	2^2	3^2	2^2	$(3:1)^2=$ 9: 3: 3: 1
3	2^3	3^3	2^3	$(3:1)^3=$ 27: 9: 9: 9: 3:3:3:1
4	2^4	3^4	2^4	$(3:1)^4$
n	2^n	3^n	2^n	$(3:1)^n$

In breeding work by crossing, therefore, if the phenotypes in the F_1 generation are in the ratio of 3 : 1 it is assumed that only *one* pair of parental characters is present; if in the ratio of 9 : 3 : : 3 : 1, or in 9 : 7 or 15 : 1. *two* pairs of characters are present; and if in the ratio of 27 : 9 : 9 : 9 : 3 3 : 3: 1, or 63 : 1, *three* pairs of characters are present.

Conception of the Germ Cells

The results of Mendel's experiments and those of later investigators lead to the following conceptions of the germ-cells:

1. Their purity with regard to the characters they bear, and their presence in equal numbers.
2. Each contains a complete set of characters (or factors that determine the characters) of the individual.
3. The fertilized egg (*zygote*) is a double structure, containing the character derived from the male parent, also that from the female. These characters may be similar (*homozygous*) or they may be dissimilar (*heterozygous*).
4. The germ-plasm of the germ-cells has a definite structure (see also page 109).

(c)—Additional Examples of Simple Mendelian Hybrids

The following examples not only illustrate the application of Mendelism to the production of new varieties of plants and animals, but also unravel some of the perplexing problems of the breeder.

1—Maize

East and Shull[1] crossed pure strains of Rhode Island White Cap, a starchy corn, and Crosby sweet corn. The F_1 generation was starchy showing the dominance of this character, but the F_2 generation showed segregation of starchy and sweet grains in the ratio of 3 to 1. When the F_1 were "selfed" the sweet grains produced sweet and 2|3 of the starchy grains produced ears containing both starchy and sweet grains in the ratio of 3 to 1, and 1-3 produced cobs producing starchy grains only. The results may be represented as follows:

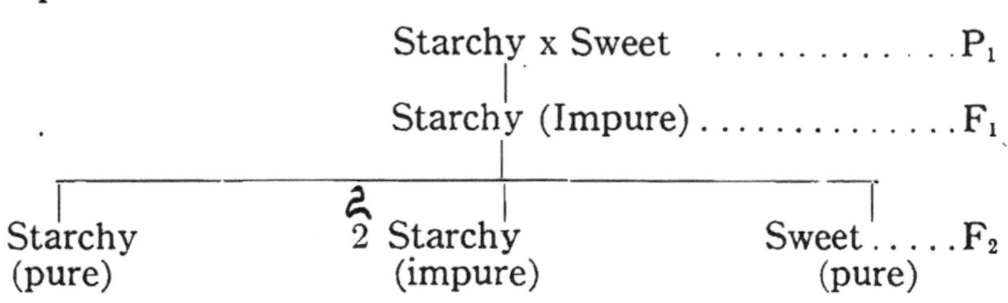

Similarly, these investigators found that when a purple corn was crossed with a white corn, the F_1 grains are purple and the F_2 grains develop in the ratio of three purple to one white.

Further they found that when a purple-sweet corn is crossed with a white starchy corn, the F_1 grains are purple-starchy, and the F_2 grains are in the ratio of nine purple-starchy grains, three purple sweet, three white-starchy, and one white-sweet—a ratio characteristic of dihybrids. The actual numbers were 1861–614–548–217.

These same investigators discovered, however, an anomaly when they crossed flint with floury corn. When the flint corn is used as the female, the F_1 grains are flinty, but when it is used as the male the F_1 grains are floury, i.e., they invariably show the character of the female. Moreover the F_2 grains develop in the ratio of one flinty to one floury. (See explanation in Babcock and Clausen, and page 124.)

(1)—Dr. E. M. East, Professor of Experimental Plant Morphology, Harvard University, was educated at the Illinois State University, where he conducted some excellent researches especially in corn breeding and genetics.

Dr. G. H. Shull (1874-), Professor of Botany, Princeton University, was educated at Antioch College and Chicago University. He served for several years in the U. S. Bureau of Forestry and Plant Industry, and in the Carnegie Institute as botanical investigator. His contributions to plant genetics have been valuable.

Primrose Hybrids.—When the "Chinese" form (*Primula sinensis*) with wavy, crenate petals and the "Star" form (*P. stellata*) with flat, entire petals are crossed the hybrids or F_1 are intermediate in form. The F_2 generation has 25% Chinese, 25% Star and 50% intermediate forms.

2—Peaches and Nectarines

Darwin many years ago showed that nectarines may come from the seeds of peaches, that peaches may come from the seeds of nectarines and that peach trees made by bud sports produce nectarines.

The explanation of these facts is now made clear since Mendel's discoveries, if we assume, as we are correct in doing, that the nectarine is a smooth variety of the peach, and that this smoothness is recessive to the hoariness of the peach. When, therefore, nectarines spring from the seeds of peaches it is a case of a recessive form arising from a dominant hybrid. When peaches are produced from the seeds of nectarines it shows that the nectarine has been pollinated from a peach.

Crossing in Oats.—Surface made crosses of strains of wild oat (*Avena fatua*) and cultivated oat (*A. sativa*) to dedetermine the inheritance of the *callus* at the base of the fertile floret.

The F_1 generation showed an intermediate form of callus, but in subsequent generations typical Mendelian segregation occurred. It was found, moreover, that the presence or absence of awns and pubescence in the hybrids was perfectly correlated with the condition of the callus as in the original forms.

Problem.—Determine the F_1, F_2 and F_3 generations in the case of crossing a wheat with a long lax head with one having a short dense head.

3—Pin-eyed and Thrum-eyed Primroses

Darwin called attention to the beautiful adaptation in these dimorphic forms of Primrose to insure cross-fertilization. Morgan (*Evolution and Adaptation, p. 369*) says: "If these two forms should appear as mutations and if, as is the case, they do not blend when crossed, but are equally inherited, they would both continue to exist as we find them to-day."

Darwin also found a third, a large-eyed form with short style and stamens low in the tube, which he termed the *homostyle* form. When Bateson and Gregory crossed this form with a small eyed thrum form the progeny (F_1) were all small-eyed short -styled, i.e. small-eye (s) is dominant to large-eye (L), and Thrum, T, (short style) to Pin, P. (long style); but the " F_2 generation consisted of four types, viz., short-sytled with small-eyes, short-styled with large-eyes, long-styled with small-eyes and homostyled with large eyes. The proportions in which these four types appeared, namely 9:3:3:1, show that the presence or absence of but two factors is concerned." Moreover, it suggests that the short-style of the homostyle form is "potentially long-styled."

The gametic formula for the F_2 generation may be resented as follows:

Female Gametes	TS	TL	PS	PL = male gametes.
TS	TSTS	TTLS	TPSS	TPLS
TL	TTLS	TTLL	TPLS	TPLL
PS	TPSS	TPLS	PPSS	PPLS
PL	TPLS	TPLL	PPLS	PPLL

Result:—9 Thrum small-eyed forms
3 Thrum large-eyed forms
3 Pin small-eyed forms
1 Pin large-eyed, which becomes the homostyled form.

(Bateson, *Principles of Heredity, p. 70*).

Castle has determined that the grey color of the wild mouse depends on the co-operation of at least six different factors, and that the "agouti" color of the wild rabbit appears on crossing a yellow with a Himalayan variety. The grey color of the ordinary rabbit depends on two factors. Other examples are the commercial carnation, a cross between the single type and the double bull-head; and Emerson's crosses with maize, gourds and beans.

Early falling of ripe grain from heads.—
Mr. Ubisch finds that this character appears in barley as the result of the presence of two Mendelian factors. When either, or both, is absent the defect does not appear.

Human Characters.—Explain why blue-eyed parents have no brown eyed-children?

4—Colors in Live Stock

The segregation occurring in the F_2 generation into pure dominant, impure dominant and pure recessive explains some of the divergent experiences of stock-breeders. For example, Black Aberdeen-Angus cattle sometimes have red calves; white-faced Herefords color-faced calves; and black French Canadian Cattle calves with a white splash; but chestnut Suffolk mares always have chestnut foals. The explanation is that the black of the Aberdeen-Angus, and the white face of the Hereford are dominants, while the chestnut of the Suffolk horse is recessive. Inasmuch, however, as all breeds are impure at their origin, impure dominants will likely persist as they are indistinguishable from the pure dominants, and recessives will occasionally appear. On the other hand ,when the recessive color is chosen as the type, as chestnut in the Suffolk horse, it is pure and will always breed true to color.[1]

It may be observed that the laws of inheritance of such unit-characters as size, shape, fertility, vigor, endurance, milk production etc., have not yet been determined. It is likely that in these cases many factors are involved.

Prof. Wilson of Dublin states that white, black, red, light dun and brown are true-breeding colors in cattle, and that yellow, dun, brindle, blue roan and red roan are hybrids and do not breed true. Of the true-breeding colors "two only behave as dominant and recessive—black being dominant to red. The others produce intermediate hybrids" as follows:

	Black	Red	Light Dun	Brown	White
Red	Black	Red	Yellow	Brindle	Red roan
Light Dun	Dun	Yellow	Light dun	Brindle	?
Brown	Brindle	Brindle	Brindle	Brown	?
White	Blue Roan	Red Roan	?	?	White

The colors of horses may be arranged in a series—"grey, dun, bay, black, chestnut, in which those coming first are dominant to all coming later, and those coming later recessive to all before them" (Wilson).

(1)—The reader should consult "Genetics in Relation to Agricuture" by Babcock and Clausen for a fuller discussion of the coat colors of domesticated animals.

No intermediate hybrids are produced. Regarding the shade factors in most of the colors, it is impossible at present to determine their inheritance.[1]

(d)—Abnormal Cases of Mendelism

In the foregoing examples the factors representing the characters of zygotes are independent and normally distributed. From the outset abnormal distributions were observed and these were looked upon as cases that did not follow the laws of Mendel. Later, however, these apparent exceptions were explained and they too fell into line in harmony with the Mendelian laws.

The disturbing causes may be grouped as follows:—

1. the absence of or imperfect dominance;
2. the presence of *multiple factors* when several different factors produce similar effects;
3. the presence of *compound factors* when certain factors interfere, or do not act unless aided by others;
4. *Linking* of two or more factors, and *crossing-over*.

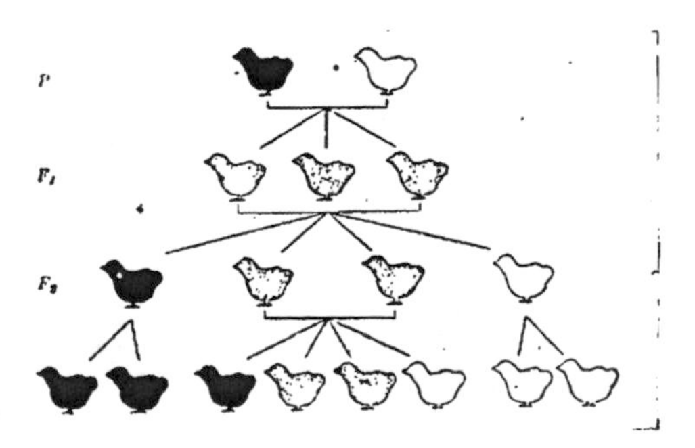

Fig. 17.—Diagram showing the course of color heredity in the Andalusian fowl, in which one color does not completely dominate another. P parental generation. The offspring of this cross constitute F1, the first filial or hybrid generation. F2, the second filial generation. Bottom row, third filial generation.

1—Imperfect Dominance

Sometimes the offspring (F_1) are unlike either parent; in such cases the dominance is said to be *imperfect*. A classic example is the Blue Andalusian fowl, which is heterozygous. The homozygous forms are the *black* and

(1)—For other theories of coat color see Babcock and Clausen, Chap. XXIX.

the *splashed-white* Andalusian. When these are crossed the progeny are *blues*, and when the blues are mated 25 per cent. of the progeny are black, 25 per cent. are splashed-white and 50 per cent. blues. (Fig 17).

The following diagram shows the inheritance—

Black x White . . . . P_1

Blue x Blue . . . . F_1

1 Black 2 Blue 1 White . . F_2

The roan color of Shorthorn cattle is produced by crossing white and red. Roan is heterozygous, for when roan is mated with roan, of the progeny 25 per cent. are red, 25 per cent. white and 50 per cent. roan.

The following diagram shows the roan inheritance:—

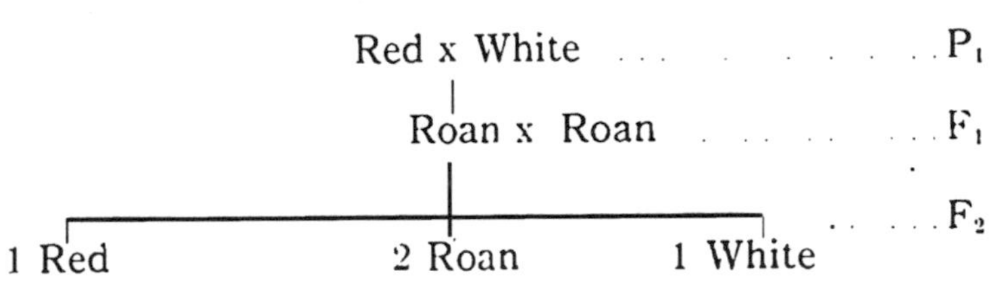

2—Multiple Factors

"The term 'multiple factors' has come, in practice, to be applied usually to cases of crossing in which two or more factor-differences occur, all of which produce similar effects" (Morgan).

Several examples of this type of inheritance have been brought to light by Nilsson-Ehle of Sweden, and by Shull, Castle and East of the United States.

(a) Nilsson-Ehle found that when a dark-brown oat with dark glumes was crossed with a white-glumed oat, F_1 were all dark brown, but F_2 were 9 dark brown, 6 light brown and 1 white. This result is explained on the assumption that the dark brown oat has two dominant factors (A and B) for dark glumes and the white-glumed oat two recessive allelomorphic factors (a and b) for light color. The gametic segregation in F_2 may be represented as follows:—

	AB	Ab	aB	ab	= male gametes
Female gametes AB	AABB	AABb	AaBb	ABab	= 9 dark brown, 6 light brown and 1 white.
Ab	AABb	AAbb	ABab	Aabb	
aB	AaBB	AaBb	aaBB	aaBb	
ab	AAab	Aabb	aaBb	aabb	

i.e.—The presence of both factors A and B in the same plant has a cumulative effect with regard to intensity of color.

(b) When a certain brown-chaffed wheat was crossed with a white-chaffed strain F_1 gave brown-chaffed, but F_2 gave 15 brown-chaffed and 1 white-chaffed. Here again there are two dominant factors for brown and two recessive factors for white. Make out a diagram after the manner in (a).

(c) When a certain red-seeded wheat was crossed with a white-seeded strain F_1 were red, but F_2 gave 63 red and 1 white. In this case there are three dominant factors for the red and three recessive allelomorphic factors for white. Make out a diagram.

Shull furnishes an interesting case where a triangular-capsuled Shepherd's Purse was crossed with a round-capsuled form. F_1 gave all triangular capsules, but F_2 gave 15 triangular and 1 round. Make out a diagram on the supposition that the plants with the top-shaped capsules have two recessive factors and have the genetic constitution *ccdd* and the pure plants with triangular-shaped capsules the genetic constitution *CCDD*.

Problem.—Analyse the following results: the F_2 ratio when certain black beans are crossed with certain white beans is 12 black : 3 yellow : 1 white.

Castle's experiments with crosses of *lop-eared* and *short-eared* rabbits show that probably four or more factors for ear-length occur hence grand-parental recessive types are not likely to appear in breeding, and that all gradations of ear-length would occur.

East's experiments with corn show that *length of ear*, *height of plant* and *productivity* depend on multiple factors.

Mendelists are inclined to believe that skin color as a result of crosses between negroes and whites is not a case of Blended inheritance but of Mendelian segregation where several multiple factors for color are concerned. (C. B. Davenport, *Heredity of Skin Color in Negro-White Crosses*, No. 188 Carnegie Inst. Wash.)

This belief is based on the fact that the color of the F_2 (offspring of mulattoes) is much more variable than that of F_1, and on the fact that white and black appear in the F_2 generation, breeding true to their respective color.

3—Compound Factors

Some of the most complex cases of Mendelian inheritance occur when certain determiners interfere with the actions of others; or do not act unless aided by others.

Four types of such compound determiners have been observed:—

(a) *The Complementary type* where the two unlike determiners become separated in breeding, and the character disappears, reappearing when the two determiners are again brought together. Examples of this type are Bateson's White Sweet Peas (Emily Henderson), East's purple maize, and Combs of fowls.

"*Emily Henderson*" *Sweet Peas.*—Bateson crossed two strains of this variety and there appeared purples. When these F_1 purples were selfed the progeny consists of whites and purples in the ratio of 7 to 9. This case of *reversion* explains the appearance of individuals with colored flowers of different types among the descendants of two white sweet peas of separate strains. The results are explained on the following assumption:—

C = Factor for color production (sensitizer).
c = Absence of factor for color production.
P = Factor for purple (basis).
p = Absence of factor for purple.

Only when the factors C and P come together will the purple color be produced.

(1) Hybrids of white varieties in the F_1 generation will have the formula CcPp—purple.

(2). Hybrids of the F_2 generation:

Female Gametes	CP	Cp	cP	cp = male gametes
CP	CCPP	CcPp	CcPP	CcPp
Cp	CCPp	CCpp	CcPp	Ccpp
cP	CcPP	CcPp	ccPP	ccPp
cp	CcPp	Ccpp	ccPp	ccpp

Phenotypically 9 purple and 7 white.

Types of Comb in Fowls.—Five types of comb in fowls exist, viz., the *single* (found in wild jungle fowl) the *rose*, the *pea*, the *walnut*, and the *breda*. Crossing the types gives the following results which are explained by Bateson and Punnett on the assumption that the Rose comb is due to the presence of a factor R dominant to an absence r, and the Pea comb to an independent factor P dominant to an ab sence factor p. When both R and P factors are present in the cells of the zygote a Walnut comb is produced, and when r and p, their recessives or absences, a Single comb results. For example the genetic formula RRPP, RrPp or RrPP represents a walnut comb, RRpp or Rrpp a rose, rrpp a single, rrPp a pea. (Fig. 18).

Fig 18.—Cross between pea and rose combed fowls: a, pea comb; b, rose comb; c, walunt comb; d, single comb.

(1) F_1 of Single x Rose gives Rose, and F_2 gives 3 Rose and 1 Single.

Female Gametes	rp	rp	= male gametes
Rp	Rrpp	Rrpp	
Rp	Rrpp	Rrpp	= all Rose

The diagram of F_2 would be:

Female Gametes	Rp	rp	= male gametes
Rp	RRpp	Rrpp	= 3 Rose and 1 Single
rp	Rrpp	rrpp	

(2) F_1 of Single x Pea gives Pea, and F_2 gives 3 Pea and 1 Single.

		rp	rp	= male gametes
Female Gametes	rP	rrPp	rrPp	
	rP	rrPp	rrPp	= all Pea

The diagram of F_2 would be:

		rP	rp	= male gametes
Female gametes	rP	rrPP	rrPp	
	rp	rrPp	rrpP	= 3 Pea, 1 Single

(3) F_1 of Rose x Pea gives Walnut and the F_2 gives 9 Walnut, 3 Rose, 3 Pea, and 1 Single.

		Rp	Rp	= male gametes
Female Gametes	rP	RrPP	RrPp	
	rP	RrPp	RrPp	= all Walnut

The diagram of F_2 would be:

		RP	Rp	rP	rp	= male gamete
Female Gametes	RP	RRPP	RRPp	RRPp	RrPp	= 9 Walnut 3 Rose, 3 Pea, 1 Single.
	Rp	RRPp	RRpp	RrPp	Rrpp	
	rP	RrPp	Rrpp	rrPP	rrPp	
	rp	RrPp	Rrpp	rrpP	rrpp	(Fig. 18)

(4) F_1 of Single x Walnut gives Walnut; the F_2 gives as in (3).

		rp	rp	= male gametes
Female Gametes	RP	RrPp	RrPp	
	RP	RrPp	RrPp	= all Walnut

(5) F_1 of Rose x Walnut gives Walnut; F_2 as in (3)

		Rp	Rp	= male gametes
Female Gametes	RP	RRPp	RRPp	
	RP	RRPp	RRPp	= all Walnut

(6) F_1 of Pea and Walnut gives Walnut; F_2 as in (3)

		rP	rP	= male gametes
Female Gametes	RP	RrPP	RrPP	
	RP	RrPP	RrPP	= all Walnut

(b) The *Supplementary* type: when two unlike determiners come together a modification occurs in the character; seen in Castle's agouti guinea-pigs;

In explanation of the results of experiments with guinea pigs Castle postulates the action of a *pattern factor* in addition to the pigment factors (See *Heredity in Coat Color in Guinea pigs and Rabbits:* Carn. Publ. 23. 1905).

Problem.—Explain the following result: 9 agouti, 3 black, 4 white in the F_2 generation when a black rabbit was crossed with a white variety with factor for making the hair yellow.

(c) The *Cumulative* type: when two like determiners on coming together effect the expression of character; observed in Miss Durham's Intensified mice;

Experiments were conducted upon mice of five types of color, viz., *albino, silver-fawn, chocolate, blue* and *black*. In explanation of the results it was found necessary to assume the presence of an *intensifying factor*, or its absence a *diluting factor*, in addition to the pigment and color factors.

(See Bateson, *Mendel's Principles of Heredity.* pp. 80-83).

(d) The *Inhibitory* type: one determiner interfering with the action of some other determiner.

Examples of such *inhibitor* factors are found when the Brown Leghorn is crossed with the Silky fowl, and when the White Leghorn is crossed with the Brown Leghorn.

(See Punnett, *Mendelism*).

Problem.—Analyse and explain the following result: 13 white: 3 colored in F_2 generation where a certain white fowl is crossed with a certain colored fowl.

In the White Leghorn the white behaves as a dominant to color, but in the White Dorking and other white birds white is recessive to color. When two such white birds are crossed the results are as given below when the following factors are assumed:

C = Factor for color.
c = Absence of factor for color.
I = Inhibitor color factor.
i = Absence of inhibitor factor.

The White Leghorn will have the gametic formula of *CI.* and the White Dorking *ci.*

The F_2 generation:—

		CI	Ci	cI	ci	= male gametes
Female Gametes	CI	CCII	CcIi	CcII	CcIi	= 12 white (flecked).
	Ci	CCIi	CCii	CcIi	Ccii	
	cI	CcII	CcIi	ccII	ccIi	3 colored
	ci	CcIi	Ccii	ccIi	ccii	1 white.

(See Punnett's *Mendelism*).

Morgan has shown in the case of *Drosophila* that the presence of certain specific factors produces sterility, and Bateson has found a definite factor for pollen atrophy in sweet peas, which behaves in a Mendelian manner.

With regard to self-sterility in beans Belling has proposed a Mendelian explanation (*Journal of Heredity*, 1914, 1916). When he crossed two fertile races of beans, Florida Velvet and Lyon, the hybrids were semi-sterile. In the F_2 generation, however, one-half of the plants had perfect pollen and ovules, the other one-half had a mixture of equal numbers of good and bad pollen grains and ovules. In the F_3 generation the progeny of the fertile plants had good pollen and ovules, but that of the semi-sterile was like F_2.

Belling proposes the presence (or absence) of a fertility (or sterility) factor A, a in race X and another similar factor B, b in race Y.

If the Velvet Bean possesses the factors AAbb, and the Lyon Bean the factors aaBB where A and B are factors for fertility and a and b are factors for sterility, the gametes for the former will be Ab and for the latter aB. The F_1 generation will have the factor-constituents AaBb which render it semi-sterile. The gametes of the F_1 generation will be AB, Ab, aB, ab, but if AB and ab are lethal to pollen grains and ovules the F_2 generation will be represented by the following diagram.

	Aa	aB
Aa	Aa Aa (fertile)	aB Aa (semi-sterile)
aB	Aa aB (semi-sterile)	aB aB (fertile)

or 50% fertile and 50% semi-sterile.

Species Hybrids.—Goodspeed and Clausen of the Cal. Exp. St. have formed a definite hypothesis for the interpretation of such phenomena as the "mutation" behavior of *Oenothera*, bud variation in horticultural plants, and the complex relations exhibited in species hybridization. The main points of the hypothesis may be outlined as follows:

a. The hereditary units of species form complex but harmonious reaction systems, the specific elements of which are the factors of Mendelian heredity.

b. Strict Mendelian inheritance follows hybridization between forms which have fundamentally like reaction-systems. The Mendelian behavior is a consequence of the existence of specific differences in certain elements in the common reaction-system.

c. When fundamentally different, but equivalent, reaction systems-are contrasted, as in some species hybrids, the elements of the two systems may not be interchangeable as in Mendelian hybrids because the diverse elements may form inharmonious reaction-systems. These inharmonious systems either may not function at all or may function in an abnormal fashion. The results of this difficulty are expressed in a variety of ways, particularly in the marked sterility of species hybrids and in the production of abnormal individuals.

4.—Linkage and Crossing-over

It will be observed after a study of the preceding abnormal cases of crossing that Mendelian views have changed with the progress of investigations, especially after the discovery that both the factors and the chromosomes undergo segregation and show parallelism in their methods of distribution. Determiners have taken the place of factors or characters, each trait being represented in the germ plasm by its own determiner.

With the more extended investigations, it became necessary to evolve the *factorial* hypothesis to explain the results. According to this hypothesis each visible character is due to the action of a large number of factors or genes in the germ plasm, each factor in turn influencing a large number of other traits. Moreover, these factors or genes are linked together in groups, in chromosomes where they are arranged in a linear series, but sometimes changing places by "crossing-over". Besides each factor or gene has a definite location in its chromosome, in a chromomere. As each chromosome carries several factors it is to be expected that these factors will remain together, even linked in groups, in the inheritance. Such linkages have been discovered. Bateson and Punnett in 1906 found that when a sweet pea with factors for purple flowers and long pollen grains was crossed with a sweet pea with factors for red flowers and round pollen grains, the two factors that came from the same parent tended to be inherited together. Later, Emerson found coupling of determiners and departures from normal Mendelian ratios when corn with red grains and white cobs was crossed with corn with white grains and red cobs. To Morgan[1] and his associates, however, we are indebted for exact, elaborate, and suggestive investigations with *Drosophila*, a fruit fly, in which linkage relations were determined.

Morgan[1] found that in Drosophila, which was normally red-eyed, there appeared in the course of breeding experiments as many as 25 distinct mutations in this eye-color. He supposes, therefore, that at least 25 factors are concerned in the production of this red eye, and that when a single one changes a different color is obtained. This one factor, however, may be called the unit factor for this particular color, so it may be treated as a simple Mendelian factor in segregation.

Again, the idea of sex-linked and sex-limited characters has been developed, with the result that many problems, formerly too intricate for solution, are now yielding to analysis.

Linkage has been shown to occur in sweet peas, primrose, snapdragon, groundsel, corn, tomatoes, wheat, oats, evening primrose, grouse-locust, bat, silk-worm, and poulttry (Morgan).

Cases of Sex-Linked Inheritance (See also page 142). Certain characters are sex-linked in that they are transmitted by individuals of one sex almost exclusively to offspring of the other sex. Color-blindness in man is an example. Men cannot hand on the character without having it, whilst women can. "A woman is only color-blind when she is homozygous for the character; a man is color-blind both when he is homozygous and heterozygous for the character. The normal women who transmit it are heterozygous for it" (Darbishire).

(1)—Prof. T. H. Morgan was born in 1866, and educated at Kentucky State College and Johns Hopkins Univ. He is now Professor of Experimental Zoology, Columbia University. His publications are "Evolution and Adaptation," "Heredity and Sex," "The Mechanism of Mendelian Heredity," and "A Critique of the Theory of Evolution."

(1) Female homozygous for a sex factor

The X Y type of inheritance. The gametic unions in color-blindness in man may be represented as follows:

Let C = Factor for color blindness.
c = Factor for absence of color blindness.
X and X = sex factors in female.
X and Y = sex factors in male.

It is obvious that each person will have two kinds of gametes with regard to color-blindness.

A normal male will have cX and Y gametes;
a normal female will have cX and cX;
a color-blind male will have CX and Y;
a color-blind female will have CX and CX and;
a carrier-female will have CX and cX.
The X chromosome carries the factor for color-blindness

The variations from equality in the sex ratio may be accounted for by one or more of the following causes: 1. the spermatozoa are of two sizes, as shown by Wodsedalek for the horse and the pig, the larger and less active with the accessory chromosome being the female-producing; the smaller and more active being the male-producing; 2. the different chromatin content of the spermatozoa may have a bearing on their mortality. (See page 143).

The gamete unions in color-blindness may be represented as follows:—

Case 1. Color-blind male x normal female.

		CX	Y = male gametes
Female Gametes	CX	cXCX carrier female	(cX)Y normal male
	cX	cXCX carrier female	(cX)Y normal male

i.e. In the children the normal males are as frequent as carrier females.

Case II. Normal male x carrier female.

		cX	Y = male gametes
Female Gametes	CX	CXcX carrier female	(CX)Y col.blind male
	cX	normal female	normal male

i.e. Half of the males are color-blind, half normal, half of the females are normal, half carriers.

Case III. Color-blind male x carrier female.

		CX	Y
Female Gametes	CX	CXCX col. bl'd female	(CX)Y col. bl'd male
	cX	CXcX Carrier female	(cX)Y Normal male

= male gametes

i.e. Half the males are color-blind and half normal. Half the females are color-blind and half carriers.

Case IV. Normal males and color-blind females.

		cX	Y
Female Gametes	CX	CXcX carrier female	CXY col.-bl'd male
	CX	CXcX Carrier female	CXY col.-bl'd male

= male gametes

i.e. The females are carriers and the males color-blind.

Case V. Color-blind male and color-blind female.

		CX	Y
Female Gametes	CX	CXCX col.-bl'd female	CXY col.-bl'd male
	CX	CXCX col.-bl'd female	CXY col.-bl'd male

= male gametes

i.e. All are color-blind.

Drosophila Hybrids (See also page 120)

Morgan's investigations on the inheritance of characters in *Drosophila ampelophila*, the Pomace Fly, furnish very complete records of sex-limited inheritance. This fly has red eyes normally but a few white-eyed forms were observed and used in crossing experiments. The following examples illustrate some of the simpler cases of inheritance. The results are explained on the assumption of the presence of the factor for Red-eye (R), the absence of this factor (r), and the sex factors as in the case of color-blindness in man. The X chromosome carries the factor for color.

Case I. White-eyed male x Red-eyed female.

The male gametes are of two kinds, viz. rX and Y, while the female gametes are all RX. The matings may, then, be represented as follows:—

		rX	Y	= male gametes
Female Gametes	RX	Red-eyed female	Red-eyed male	
	RX	Red-eyed female	Red-eyed male	= all red-eyed

The diagram of the F_2 generation is—

		RX	Y	= male Gametes
Female Gametes	RX	Red-eyed female	Red-eyed male	50% red-eyed females 25% red-eyed males. 25% white-eyed males.
	rX	Red-eyed female	White-eyed male	Ratio—3 red to 1 white

Case II. White-eyed male x F_1 red-eyed female (Heterozygous).

The male gametes are of two kinds: rX and Y, while the female gametes are RX and rX. The matings may be represented as follows:—

		rX	Y	= male gametes
Female Gametes	RX	Red-eyed female	Red-eyed male	25% red-eyed females 25% red-eyed males
	rX'	White-eyed female	White-eyed male	25% white-eyed females 25% white-eyed males.

Case III. Red-eyed male x white-eyed female.

The male gametes are RX and Y, and the female gametes rX and rX.

		RX	Y	= male gametes
Female Gametes	rX	Red-eyed female	White-eyed male	red-eyed females and
	rX	Red-eyed female	White-eyed male	white-eyed males

Showing that the red-eyed male parent is heterozygous for color.

Case IV. Red-eyed male x F_1 red-eyed female (heterozygous).

The matings may be represented as follows:—

		RX	Y	= male gametes
Female Gametes	RX	Red-eyed female	Red-eyed male	Same as the F_2 page 113
	rX	Red-eyed female	White-eyed female	

Hybrids of Dorset and Suffolk Sheep

Dorset Ram (horned) x Suffolk Ewe (hornless).

The results are explained on the assumption of the following factors:—

H = factor for horns.
h = factor for absence of horns.
XX = sex factors in female; X, Y in male.
In horned ewes H & X must be duplex.
In rams H and X are simplex.

Following are the possible matings:

Case I. F_1 generation.

		HX	HY	= male gametes
Female Gametes	hX	hornless female	horned male	= 50% males horned and
	hX	hornless female	horned male	50% females hornless.

Case II. F_2 generation.

		HX	hX	HY	hY = male gametes
Female Gametes	HX	HHXX horned female	HhXX hornless female	HHXY horned male	HhXY horned male
	hX	HhXX hornless female	hhXX hornless female	HhXY horned male	hhXY hornless male

Giving 3 horned males,
3 hornless females,
1 horned female, and
1 hornless male.

(See Articles by Arkell and Davenport, *Science*, 1912, for other results).

Wentworth of Kansas Agricultural Experiment Station reports, the herd Ayrshire bull being a dark mahogany and white in color, popularly known as black, that (1) crosses with red females produced all black males and all red females; (2) the F_2 generation shows a black male and a black female; (3) crosses with black females produced one black male and three black females; (4) crossing a similar black female with a black male (heterozygous) gave 2 black males, 2 black females, and one red female.

There is here evidence of color being linked up with sex.

(2) Female Heterozygous for Sex Factor

The WZ type of inheritance

Hybrids of Abraxas Moths.

Mention has already been made of the probability that the females of *Abraxas* are heterozygous and the males homoyzgous with regard to the sex factor.

The results of crossing *A. grossulariata* and *A. lacticolor*, a pale variety, by Doncaster and Raynor are explained on two assumptions:

(1) The female is heterozygous for sex, femaleness being dominant, and the male homozygous recessive;

(2) The factor for color for grossulariata is dominant.

The gametic formulæ for the crossings are given below:

C = dominant foctor for color (*grossulariata*)
c = recessive factor for color (*lacticolor*)
WZ = sex factors in female (two kinds of ova)
WW = Sex factors in male (one kind of sperm)

Case I. Grossulariata (male) x lacticolor (female).

The male gametes are all of one kind-CW, while the female gametes are of two kinds-cW and cZ.

The matings may be represented as follows:—

		CW	CW
Female Gametes	cW	CcWW Gross. male	CcWW Gross. male
	Z	CWZ Gross. female	CWZ Gross. female

(CW CW = male gametes)

i.e. All *grossulariata*, half being male.

Case II. F_1 Grossulariata male x F_1 Grossulariata female.

		CW	cW
			= male gametes
Female Gametes	CW	CCWW Gross. male	CcWW Gross. male
	Z	CWZ Gross. female	cWZ Lacti. female

i.e 75% *grossulariata, 25% lacticolor female.*

Case III. F_1 Grossulariata (male) x lacticolor (female)

		CW	cW
			= male gametes
Female Gametes	cW	CcWW Grossu. male	ccWW Lacti. male
	Z	CWZ Grossu. (female)	cWZ Lacti. female

i.e. 50% lacticolor, 50% grossulariata.

Case IV. Lacticolor (male) x F_1 grossulariata, (female)

		cW	cW
			= male gametes
Female Gametes	CW	CcWW Grossu. male	CcWW Grossu. male
	Z	cWZ Lacti. female	cWZ Lacti. female

i.e. 50% *grossulariata* males. 50% *lacticolor* females.

Fowl Hybrids. When Barred Plymouth Rocks were crossed with Cornish Indian Game or Black Langshans, results similar to those obtained in Abraxas were obtained. The female Barred Plymouth Rock is evidently heterozygous for color.

B = dominant factor for Barring.
b = recessive factor.
WZ = sex factors in female (two kinds of ova).
WW = sex factors in male (one kind of sperm).

Case I. Barred Plymouth Rock (male) x Cornish Indian Game (female).

The matings may be represented as follows:—

Female Gametes	BW	BW
bW	BbWW Barred male	BbWW Barred male
Z	BWZ Barred female	BWZ Barred female

(BW BW = male gametes)

i.e. all barred fowls.

Fig. 19.—Results of crossing a Black Langshan (male) with a Barred Plymouth Rock (female).

Case II. F_1 Barred (male) x F_1 Barred (female) of Case I.

Female Gametes	BW	bW
BW	BBWW Barred male	BbWW Barred male
Z	BWZ Barred female	bWZ Black female

(BW bW = male gametes)

i.e. 25% black females.

Case III. Cornish Indian Game (male) x Barred Plymouth Rock (female).

		bW	bW	= male gametes
Female Gametes	BW	BbWW Barred male	BbWW Barred male	
	Z	bWZ Black female	bWZ Black female	

Case IV. F_1 Barred (male) x F_1 Unbarred (female) of Case III.

		BW	bW	= male gametes
Female Gametes	bW	BbWW Barred male	bbWW Black male	
	Z	BWZ Barred female	bWZ Black female	

i.e. 50% barred, 50% unbarred:
(See Fig. 19).

Morgan says that the results of such crossing "means that the Barred Plymouth Rock is a black race with an additional dominant factor for barring. The Black Langshan is the same black race but without the barring factor."

Egg Production in Fowls

The inheritance of egg-production or fecundity has been shown by Dr. Pearl to be sex-linked. As the subject is too intricate for full discussion in a work of this kind reference will be made only to the main features.

Pearl found (a) that high egg-production could not be obtained by selecting good-laying hens only as mothers; and (b) that the laying capacity of hens could be ascertained at an early stage by noting the first winter production. On this basis he classifies hens into three groups: (1) those laying over thirty eggs during their first winter will lay from 150–200 eggs during the year, (2) those laying fewer than thirty eggs during the winter will lay from 80–120 eggs, and (3) those laying no eggs will lay from 40–60 eggs.

With the above groups in mind Pearl advanced the working hypothesis that the egg-laying capacity of hens depends on the presence or absence of three separate factors which act in a Mendelian manner. These factors are:—1. A fecundity factor (A,a), making for a winter production of less than thirty eggs; 2. a sex-linked fecundity factor (B,b), also making for a winter egg-production of less than thirty eggs; when these two factors are joined a winter production of more than thirty eggs occurs; and 3. sex factors of the WZ type of inheritance, the female being heterozygous and the male homozygous, and the factor for femaleness being linked with the recession fecundity factor b.

Pearl's three groups of hens probably represent six genotypes: WBAZbA, WBAZba, forming the high producers of group 1; WBaZba, WbAZbA, WbAZba forming group 2, and WbaZba forming group 3.

The cocks would be divided into two groups representing nine genotypes: WBAWBA, WBAWBa in group 1, from which can be expected high producing daughters; and WBaWBa, WBAWbA, WBAWba, WBaWba, WbAWbA, WbAWba and WbaWba.

"It is, therefore, no use simply to select hens from mothers with high egg production, since the factor which determines this character is transmitted by the mother only to her male offspring." To get a strain of high producers "the father of the selected hens should be the son of a hen with high egg production, and only by mating such cocks with hens which themselves are good layers can a pure strain be built up."

(See also Babcock and Clausen, pp 560–562 and 585–590, and Wilson pp. 99–103).

What proofs (besides that mentioned above) can be given for the belief that high fecundity is inherited from the sire?

In the cases of *Drosophila*, man and sheep, it will be observed that the male transmits certain characters to his daughters only, while in the cases of *Abraxas* and Fowl the female transmits certain characters to her sons only. In the former cases the characters appear in the males and are transmitted by the females That is, if a color-blind man marries a normal woman the offspring are normal. But if a daughter marries a normal man some of the sons will be color blind.

If a white-eyed male *Drosophila* is mated to a red-eyed female the offspring are red-eyed. If these are inbred all the F_2 daughters are red-eyed, but half the sons are white-eyed. "In a word, the grandfather transmits his characters visibily to half of his grandsons but to none of his granddaughters." (Morgan).

In the second case the characters appear in the females and are transmitted by the males. That is, if a *lacticolor* (*Abraxas*) female mates with a *grossulariata* male the progeny are all *grossulariata*, but if these are inbred the male progeny are *grossulariata* and half the females *lacticolor*. "The *lacticolor* grandmother has transmitted her peculiarity *visibly* to half of her grand-daughters but to none of her sons" (Morgan).

Laboratory Exercises with Drosophila

The *Drosophila* fly is one of the most convenient organisms to employ with classes in demonstrating the nature of Mendelian inheritance. It can be readily secured about fermenting fruit, and can be reared in large numbers in sterilized vials containing bits of fermenting bananas.

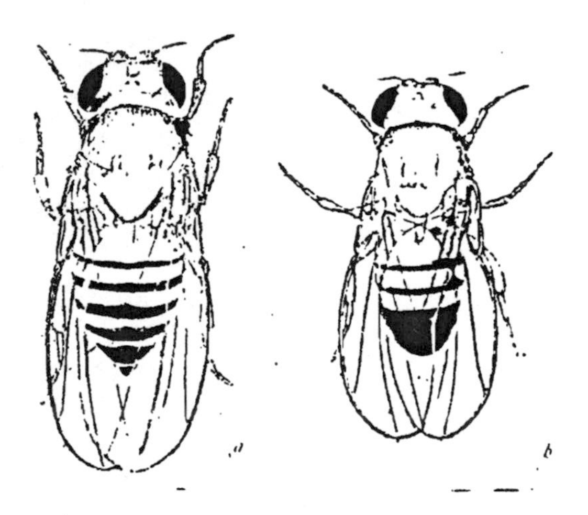

Fig. 20.—Drosophila ampelophila. a, female and b, male (After Morgan.).

Precaution must be taken however, to prevent bacterial and fungal infection of the bananas. Mutants will usually be found where a large number of the flies are bred. The normal fly has a *grayish-yellow* body, *long* wings and *red* eyes, but forms may be discovered having *black*, *yellow* or *ebony* bodies; *miniature* or *vestigial* wings; and *white* or *sepia* eyes. (Fig. 20).

With such material numerous exercises may be conducted, as follows:—

1. Cross normal with vestigial winged forms, and "carry on" through the F_2 generation.
2. Cross a normal fly with a black-bodied fly, and "carry on" through the F_2 generation.
3. Make reciprocal crossings of gray and yellow flies and "carry on" through the F_2 generation.
4. Make reciprocal crosses of a white-eyed male, if such can be found, with a red-eyed female and "carry on" through F_2

Etc.

Give the interpretation of the results. What was Morgan's method of locating the 150 distinct characters in *Drosophila* into four groups? (See Morgan's *Critique*).

Drosophila is readily and conveniently reared in bottles of banana-agar inoculated with yeast. Two or three over-ripe bananas are mashed in 200 cc water and added to 200 cc of liquid agar, made by adding 4 grams of agar to 200 cc of water and autoclaving until the agar is thoroughly dissolved. The banana-agar is then poured into 6 oz. cotton-plugged, wide-mouthed bottles to a depth of half an inch. On solidification two or three drops of yeast solution are added to each bottle. For student-work, test-tubes or vials may be used.

The sexes of *Drosophila* may be distinguished by the following characters of the abdomen:—

Male	Female
(1) Three narrow bands and a black tip.	(1) Five black bands.
(2) Black caudal band extends around the under side.	(2) Black caudal band never meets on the under surface.
(3) Caudal extremity rather round and blunt.	(3) Caudal extremity sharp and protruding.

(Consult *Genetics Laboratory Manual* by Babcock and Collins for details regarding the method of breeding and experimenting with Drosophila).

Crossing-over

Reference has already been made to the fact that the expected Mendelian ratios in segregation do not always occur. Such departures are due in many cases to linkage of factors in the same chromosome, and in some cases to "crossing-over" of factors from one of the homologous chromosomes to the other at synapsis preparatory to reduction. Linkage of characters is probably more common than sex-linkage where the characters lie in the sex chromosome.

Instances of sex-linked inheritance have already been discussed, consequently only a brief treatment will be given of linkage of other factors.

Morgan says: "When factors lie in different chromosomes they give the Mendelian expectations; but when factors lie in the same chromosome they may be said to be *linked*, and they give departures from the Mendelian ratios. For example, Morgan crossed a gray vestigial-winged male Drosophila with an ebony long-winged form. The F_1 were all gray long-winged flies, and the F_2 gave the normal Mendelian ratio of 9:3:3:1. In this case the factors lay in different chromosomes.

But when he crossed a black vestigial-winged male fly with a gray long-winged female form, the F_1 were all gray long-winged flies. When, however, it is known that the factors *black* and *vestigial* lie in the same chromosome and the factors *gray* and *long* lie in another chromosome, it will be clear that the cells of the F_1 carry two kinds of chromosomes and that its gametes should be by and *GL* for each sex. The F_2 should be expected to give black vestigial and gray long in the ratio of 1:3 according to the following diagram:

Male Gametes	bv	GL = female gametes
bv	bv GV gray vestigial	bv GL graylong
GL	GL bv graylong	GL GL graylong

Actual breeding results ,however, included in addition a small percentage of black long and gray vestigial forms. This departure is explained if the female gametes are by, GL, bL, Gv, according to the following diagram:

Male Gametes	bv	GL	bL	Gv = Female
bv	bv bv black vestigial	bv GL graylong	bv bL black long	bv Gv gray vestigial
GL	GL bv graylong	GL GL graylong	GL bL graylong	GL Gv graylong

If the female gametes had been all of the same number the ratio would have been

1 black vestigial,
5 gray long,
1 black long and
1 gray vestigial,

but as the two latter appeared in about 17% of the cases it is evident that the gametes do not exist in equal numbers.

Instead of crossing an F_1 male with an F_1 female. the crossing-over can be more readily made out by back-cross-

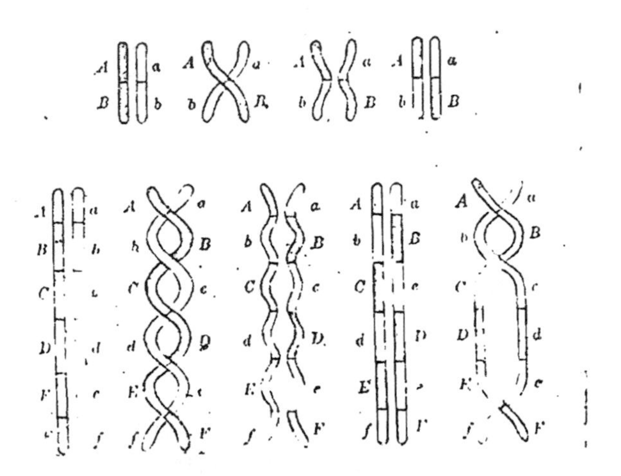

Fig. 21.—Diagram of the Probable Mechanism of "Crossing-over." Pairs of homologous chromosomes, one from the father and the other from the mother, are shown in synapsis; on the left they lie parallel to each other and when they separate they remain as they were before union; in the second column they are shown crossing each other one or more times; in the remaining figures are shown the results of the chromosomes breaking at the points of crossing, thus interchanging sections of the two chromosomes. Letters indicate loci of homologous genes in the two chromosomes of a pair. (After Wilson.).

ing an F_1 female to a black-vestigial male. Four kinds of flies are produced, viz., black-vestigial, gray long, black-long and gray vestigial, according to the following diagram:

		bv	GL	bL	Gv	= female gametes
Male Gametes	bv	bv bv black vestigial	GL bv gray long	BL bv black long	Gv bv gray vestigial	

The production of such gametes is explained by Morgan on the theory of *crossing-over*. At synapsis the two chromosomes, one containing *bv*, the other *GL*, sometimes become slightly twisted so that they do not separate completely but an exchange or crossing over of factors takes place with the formation of two mixed chromosomes (Fig. 21). Morgan states that this crossing-over in the F_1 females occurs in 17% of black vestigial and gray long flies. The effect of this 17% of crossing-over on the F_2 is to produce about 21% black vestigial, 70% gray long, 4% black long, and 4% gray vestigial. (Fig. 21).

Students who may wish to pursue this phase of inheritance further are referred to Morgan's *Mechanism of Mendelian Heredity*, and *Physical Basis of Heredity*.

(e)—Endosperm Inheritance

Reference has been made to the phenomenon of *Xenia* in corn (See page 30). We know that the endosperm is not a part of the embryo. It is the product of the fusion of a male cell with a fusion nucleus which is formed by two nuclei. It is, therefore, a hybrid of three nuclei (a triple fusion), and appears the same year the seed is planted.

The crosses of sweet and starchy corn, and red and white corn, show typical xenia as it was first understood, i.e. the direct effect of foreign pollen on the endosperm.

An interesting result occurs, however, when the pollen of a white corn falls on the silks of a red corn, for a red corn is produced. In this case the color comes not from the pollen, but from the fusion nucleus of the red corn. The results of these reciprocal crosses may be explained as follows:—R and W are the two factors for color, R dominant; In the first case the endosperm has the genetic female of R WW, in the reciprocal WRR. Both will, therefore, be red.

The above illustration and the one given on page 30 are sometimes cited as cases of *maternal inheritance*.

(f)—Breeding for Quality and Quantity of Milk

Improvement of Quality of Cow's Milk.

We must look to Denmark for an example of such improvement, for there milk records have been kept for nearly fifty years. The improvement is to be attributed (1) to the use of sires who were the sons of cows giving high quality milk and who also produced daughters giving high quality milk; and (2) to the breeding from dams giving high quality milk.

The records of the herds studied show clearly that the quality of the milk depends on the dam as well as the sire. It is impossible as yet to determine the Mendelian factors concerned, but by selection according to milk records, and by mating the highest, animals are produced with all the factors, whatever they may be, necessary for the production of high quality milk.

Breeding for Milk Production.

Few records have as yet been published regarding the inheritance of quantity in milk, but these seem to show that quantity and quality are inherited separately. It must be borne in mind that milk yields fluctuate with food, weather, comfort, shelter, health, age and lactation period.

Professor Wilson of Dublin suggests a method of calculating the milk yield which is free from many of the uncertainties referred to. He says that about half the cow's total yield for a normal lactation is given by about the end of three months. This part is freer from variation than the remaining part ,and may be used to determine the grade of the cow. He computes that a cow giving 5½ gals. a day at her maximum will give over 1000 gallons during a normal lactation; a 4-gallon cow will give 800 gallons, and a 2½-gallon cow about 500 gallons. That is, the maximum when multiplied by 200 gives the total yield approximately.[1] To breed up a milking herd, first secure a high grade bull, i.e. a bull which has produced all high grade daughters from high grade cows, or half his daughters high grade by medium grade cows, or all his daughters medium from low grade cows. By continuing the application of this principle it is possible to grade up a herd in time so that all its members will be high grade.

(g)—Atavism and Reversion

When a form arises which resembles a grandparent in some characters rather than a parent, the term *atavism* has been given to designate the occurrence. When a form possesses a character which does not appear in any near generation the case is one of *reversion.*

Atavism is often readily explained by Mendelian laws. For example, a blue-eyed child of heterozygous brown-eyed parents has two blue-eyed grandparents. This is a case of

(1)—Canadian and American authorities are of the opinion that this method of computing the milk yield is not sufficiently accurate for scientific purposes. They prefer to trust to the systematic weighing of the milk and to reckon the total milk yield from the daily or weekly records.

atavism, but it is readily explained by Mendelian laws, as follows:—

Brown-eyed x blue-eyed   Blue-eyed x brown-eyed.... P_1

Brown-eyed (impure)   x   Brown-eyed (impure)........ F_1

Brown-eyed (pure)   Brown-eyed (impure)   Blue-eyed (pure)...... F_2

Reversion is also sometimes explained by Mendelian laws as latent characters appearing on segregation. Examples occur when varieties of white -flowered sweet peas are crossed. (See Chapter 15). The progeny are purple like the wild form; also when the Bush sweet pea is crossed with the dwarf Cupid sweet pea (Punnett). Foals sometimes have a few stripes on the fore legs, and slaty-blue pigeons occur among buff and white domestic pigeons. Read Darbishire's account of the results of crossing Albino mice with Japanese waltzing mice.

Problems.—1. Account for the occasional appearance of a horned animal and of a red colored calf in herds of Aberdeen-Angus.

2. Account for the occasional appearance of a black hen in the progeny of Barred Plymouth Rocks.

(h)—Inheritance in Aphids or Plant Lice

The life history of aphids is well known. The fertilized egg hatches out a female, the *stem-mother*, the following spring, followed by a succession of generations of females. On the approach of autumn a brood of sexual males and females appears. The explanation of these phenomena is not clear, but some progress has been made in solving the mystery. Morgan has shown that the spermatids of Phylloxera are of two kinds, but those without an accessory chromosome degenerate. Consequently only those containing an accessory chromosome take part in fertilization, and the fertilized eggs produce females. The problem of the production of the males parthenogenetically, however, at the approach of autumn has not yet been solved. It is probable, however, that external or environmental factors are to some extent responsible. In *P. caryæcaulis* one-stem mother gives rise to the line ending in sexual females, and another stem-mother to the line ending in the males. On the other hand in other species of *Phylloxera* and in many aphids the same stem-mother produces both lines.

In this connection it is interesting to observe the production of males and females among bees, wasps and ants. It is well known that fertilized eggs of the queen bee produce females and unfertilized eggs males. The mature egg has one sex chromosome and the male cell has but one. When the egg is fertilized by a male cell the product has two sex chromosomes, characteristic of the cells of a female bee; when unfertilized the egg develops into a male.

Nature and Nurture Again.—"According to Mendel's law certain ancestors contribute nothing to the constitution of certain offspring in respect of certain characters" (Lock), hence all individuals of a race may and often do not possess the same characters. Neither are the differences between individuals of the same race always the result of unequal development of the characters or even of their latency.

A distinction is here made between racial and varietal characters, but no matter how it is composed "every inheritance requires an appropriate nurture if it is to realize itself in development. Nurture supplies the liberating stimuli necessary for the fuller expression of the inheritance" (Thomson).

It is often difficult to say which is the more important factor in the life of the organism—Nature or Nurture. The organism is dependent upon its surroundings for its development. In unappropriate, unfavorable surroundings the natural inheritance cannot find expression for its highest possibilities. On the other hand the highest development cannot be attained even in the best environment by an organism with poor natural inheritance (See section "*Eugenics*").

In Mendelian inheritance every character is represented by one or more determiners or factors. This idea may profitably be introduced into the discussion of Nature and Nurture. Here, however, it must be assumed that there are two sets of factors—those of environment and those of inheritance—every character being the product of these two sets of factors. If either set be altered the organism will be altered, that is, there will be variation. "Only those characters appear regularly in successive generations which depend for their development on stimuli always present in the normal environment. Others, depending on a new or occasional stimulus, do not reapppear in the next generation unless the stimulus is present" (Goodrich).

Range of Application of Mendel's Laws.—The unravelling of many intricate cases of crossing during the last few years has brought some supposed instances of Blended Inheritance into the Mendelian list (see pages 74-95). Enthusiastic Mendelists assert that the law holds true for all sorts of characters in both plants and animals, but this remains to be seen. It must be remembered, however, that many complications often arise to blur the segregation, such as *imperfect dominance, multiplicity of factors, direct effect of environment, etc.*

Some plant breeders believe that it will be possible soon to predict varieties with some degree of accuracy. Gilbert gives his reasons for doubting this belief as follows:

"(1) We do not know what plants will Mendelize until we try.

(2) Even in plants that do not Mendelize, one half of the offspring have stable characters.

(3) Mendel's laws deal primarily with mere characters not with a variety or with a plant as a whole.

(4). The breeder usually wants wholly new characters as well as recombinations of old ones, or he wants augumented characters, and these lie outside the true Mendelian categories.

(5) New and unprofitable characters are likely to arise from the influence of environment or other causes.

(6) Variability itself may be a unit character and therefore pass over.

(7) Many plants with which we must work will not close-fertilize; some are monoecious or diœcious.

(8) One can never predict just what combination of characters any plant of a cultural variety will have even though it be strictly Mendelian."

The new biological principles, especially Mendel's laws, are of value to the commercial plant breeders. They are *time-savers*, for the breeders know that it is in the second hybrid generation that segregation takes place, not in the first. Moreover, it can be now ascertained what characters are heterozygous and therefore unfixable.

It is interesting to note that Weismann and Mendel differ in their conception of the constitution of the fertilized egg. The former believed that all the characters of both parents are represented by *ids* in the fertilized egg. The latter showed that characters are segregated in the reproductive cells, only one of two contrasting characters of the parents being present.

"A reproductive cell, therefore, contains a selection of characters from its ancestry, and not a complete collection" (Coulter).

Chapter 16 — BIOLOGIC METHOD OF INVESTIGATING HEREDITY

In the preceding discussion of Mendelian inheritance, especially in connection with linkage or coupling, "crossing-over" and sex, it was found necessary to assume the existence of chromosomes, as bearers of the heredity factors. This chapter will deal more particularly with the study of chromosomes, which is the third or *Biologic method* of studying heredity. (See page 46).

(a)—The Cell in Heredity

It is known that among the simplest plants and animals every cell has the power to reproduce the entire organism. However, in the higher multi-cellular plants and animals the power of reproducing the organism is restricted to the reproductive or germ-cells. Looked at from this standpoint the germ-cells of the higher organisms are not cells endowed with special powers, but cells that have retained this primitive power, belonging to cells of the simplest plants and animals.

It should also be borne in mind that reproduction involves cell-differentiation, as well as cell-multiplication. Heredity transmission is, therefore, concerned with cell structure, and the factors that control the machinery of the cell. Parental likeness, as well as variation, is to be explained; also the appearance of new characters and the loss of old ones.

The Cell.—"A cell is a body consisting essentially of *protoplasm* in its general form, including the unmodified cytoplasm, and the specialized *nucleus* and centrosome (in animals), while as unessential accompaniments (especially in plants) may be enumerated:

The cell membrane, starch grains, pigment granules, oil globules, chlorophyll granules and protein granules. (Fig. 22).

New cells arise by a process of division from the mother cell, which divides into two cells, and each of these into two cells, and so the cells multiply until the body of the plant or animal is constructed. This original mother-cell is called the fertilized egg-cell or *zygote*, which as we have already

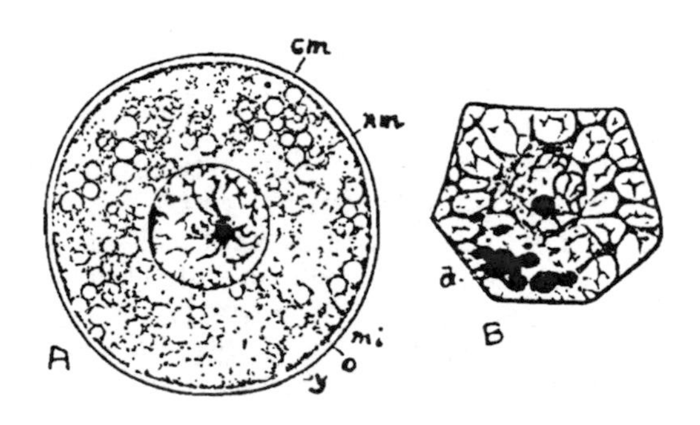

Fig. 22.—A, ovarian ovum of cat; cm, cell membrane; mi, microsomes; o, nucleolus; nm, nuclear membrane; y, yolk alveoli. B, cell from root cap of callalily; d, plastids.

noted, arises by the fusion of two independent cells—the gametes or germ-cells—male and female. The female gamete is called the egg-cell or ovum, the male gamete, the sperm-cell, spermatozoon or spermatozoid. (See also page 43).

Nuclear Division. In the division of cells the nucleus plays a prominent part. On staining the cell the nucleus is seen to have a complex structure. It is composed of a network of fibrils, enclosing in its meshes a clear substance. Embedded in the fibrils are granules of a substance which stains very deeply, called chromatin. When a cell is about to undergo division the chromatin substance changes from a network to a tangled thread which gradually shortens and thickens and breaks up into a number of rods called *chromosomes*. The nuclear membrane disappears about this time. Then appear a number of delicate fibrils, arranged in the form of a spindle, and the chromosomes arrange themselves along the equator of the spindle. They are next seen to split lengthwise into equal portions, and soon the two sets migrate towards the poles of the spindle. As they reach the poles they gradually elongate and fuse so as again to form the chromatin network of two new nuclei. A wall begins to form between the two nuclei, and soon two daughter-cells take the place of the original mother-cell. (Fig. 23).

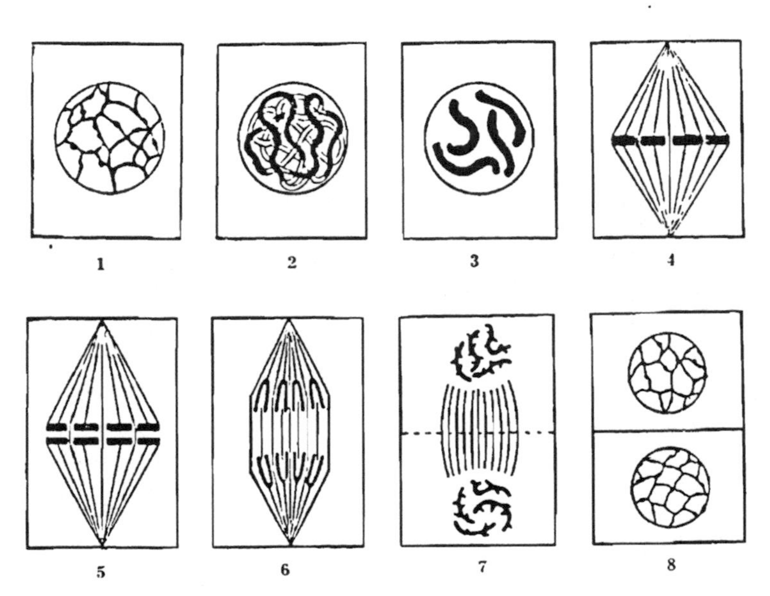

Fig. 23.—DIAGRAM OF STAGES IN NUCLEAR DIVISION. (After Lock).

1—The resting cell; 2—The chromatin forming a tangled thread; 3—Thread forming chromosomes; 4—The nuclear membrane disappears, the spindle forms and the chromosomes form the equatorial plate—this marks the end of the prophase stage; 5—Each chromosome splits lengthways— the metaphase stage; 6—Each half-chomosome moves towards the poles of the spindle; 7—The half-chromosomes come together at the poles—the end of the anophase stage; 8—The network is formed again, a wall is laid down between the two cells, and result is two cells, each with its nucleus—the end of the telophase stage.

(b)—Number of Chromosomes

The number of chromosomes varies from 2 to 200 and more in different kinds of animals and plants, but they are always constant for the same organism. It is believed by many that the chromosomes are really the transplanters, or the bearers, of hereditary characters inasmuch as they are persistent organs of the nucleus and at every division they appear in the same number. No other portion of the cell is apparently transmitted from generation to generation.

PLANTS.	*Number of Diploid or 2N Chromosomes.*
Canna indica	6
Crepis	6
Chrysanthemum	90
Lilium mastagon	24
Drosera sp. (Sundew)	20
Drosera sp	40
Onion	16
Oenothera gigas	28
" *scintillans*	15
" *lata*	15
" *biennis*	14
" *lamarckiana*	14
Pisum sativum (Garden Pea	14
Triticum sativum (Wheat)	16
Rye	16
Tomato	24
Four O'Clock	32
Corn (*Zea maidis*)	40
Tobacco	48
Cotton	56
Gymnosperms	12
Nightshade	72
ANIMALS.	
Honey bee (male)	16
Honey bee (female)	32
Drosophila ampelophila	8
Mouse	40
Salamander	24
Trout	24
Sagitta	18
Ox	16
Guinea pig	16
Man	48, 47
Grasshopper	12
Artemia	168
Ascaris	4
Leptinotarsa (Potato Beetle)	36
Snail (*Helix hortensis*)	44
Abraxas	56
Anasa tristis (Squash Bug)	22, 21

Horse	38, 36
Rabbit	42, 28-36, 22
Cat	36
Hog	20, 18
Fowl	18, 17
Rana catesbiana (Frog)	29
Nematus ribesii (Currant saw fly)	8
Diabrotica vittata (striped cucumber bettle)	21
Cricket (*G. domesticus*)	21
Gypsy Moth	31
Eristalis tenax	12
Culex pipiens (common mosquito)	6
Spittle insect (*P. spumarius*)	24
Pyrrhocoris apterus	24
Acridium granulatus	13
Protenor belfragei	14, 13
Lygaeus bicrucis	14
Oncopeltus fasciatus	16
Cicada	20, 19

(c)—Maturation and Reduction

It is known that the nucleus of the zygote or fertilized egg-cell contains twice as many chromosomes as either of the gametes. The chromosomes of the gametes at fertilization do not fuse, but remain separate. It is clear, therefore, that there must be a reduction in the number of chromosomes prior to the formation of the gametes to one-half, else the number in the zygote would double with each generation. This reduction has actually been observed.

It is now generally agreed that the reduction in the number of chromosomes in the higher animals occurs in the last two-cell divisions by which the definite germ-cells arise; that is, when the ovarian oocyte gives rise to the mature ovum and two or three polar bodies, and when a spermatocyte divides into four young spermatozoa. (See Fig. 6).

In plants, the reducing division occurs at the formation of the spores, which arise in sets of fours.

In the formation of sporangia of ferns, the central cell gives rise to *tapetum* and *archesporial cell.* The latter soon forms *sporogenous* or *grand-mother* cells of the spores. When the mother-cell divides to form four spores, the number of chromosomes is halved. This number remains con-

(1)—The recent studies of Miss Carothers of the chromosomes of certain Orthoptera (1913 and 1917) point very definitely to the conclusion that the assortment of the chromosomes at maturation is a random one, thereby bringing the chromosome theory into harmony with the Mendelian law.

stant through the sexual or gametophyte stage, which is also called the *x* generation. When fertilization occurs each gamete contributes its quota of chromosomes, thus producing the *2x* generation.

In the case of ferns, the spores germinate and give rise to prothallia; on the prothallia develop the gametes—male and female—still with the reduced number of chromosomes.

By fertilization or fusion of the gametes a zygote is produced with the normal number of chromosomes. In flowering plants the gametophyte stage is much further reduced.

The formation of the sex cells, sperms or eggs, in the higher animals may be represented by the following diagrams:

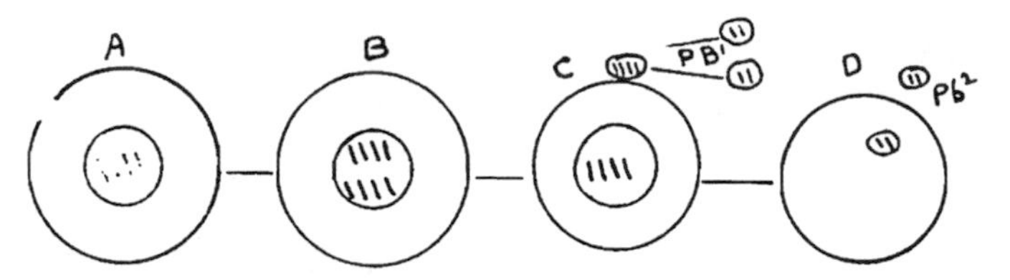

FIG. 24.

Ogenesis, Polar Bodies and Reduction: A—Grandmother egg-cell of Ascaris with four chromosomes; B—Fully developed egg mother-cell with eight chromosomes, the result first of fusion in pairs to form two bivalent chromosomes, then the formation of tetrads; C—Oocyte with four chromosomes, and formation of first polar body (Pb1) with four chromosomes, which soon divides into two polar bodies with two chromosomes each; D—Formation of second polar body (Pb2). The nucleus of the new mature ovum has two chromosomes. (After Hertwig).

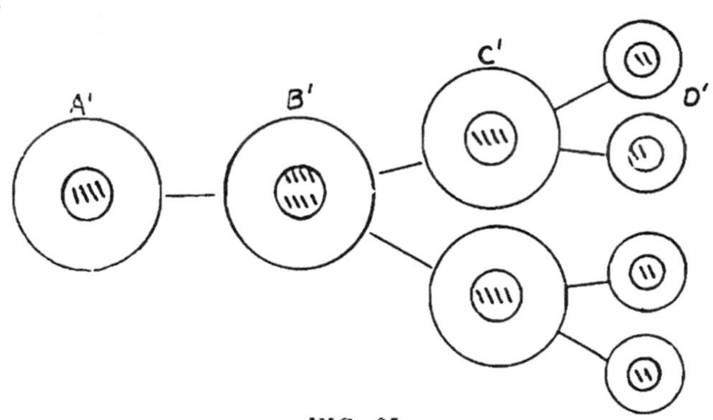

FIG. 25.

Spermatogenesis and Reduction: A—Grandmother sperm-cell of Ascaris with four chromosomes; B1, sperm mother-cell with eight chromosomes, the result first of fusion in pairs to form two bivalent chromosomes, then the formation of tetrads; C1, two spermatocytes, each with four chromosomes; D1, four spermatozoa formed each with two chromosomes. (After Hertwig).

The following diagrams (Figs. 26-27) are presented with the object of visualizing the distribution of the chromosomes in the formation of the gametes or sex-cells. In Fig. 26 the somatic and primary sex-cells contain but one pair of chromosomes (one paternal and one maternal). When the reduction process begins (synapsis) in the spermatocyte or oocyte, the chromosomes fuse (Fig. 26, 2), then they divide lengthwise and transversely and form a tetrad (Fig. 26, 3). By two rapid cell divisions four germ cells are formed, two with a paternal chromosome and two with a maternal chromosome.

In Fig. 27 each of the somatic and primary sex-cells contains two pairs of chromosomes (two paternal and two maternal). At synapsis they fuse in pairs (Fig. 27, 2), and later form two tetrads. In the two rapid cell divisions by which the germ-cells are formed, the separation of the chromosomes can take place in two ways, (Fig. 27 A. and B.) and with equal frequency. Consequently the germ cells will be of four kinds according to their chromosome content.

When three pairs of chromosomes are present in the somatic and primary sex-cells it can be readily shown in the same manner that there will be three tetrads formed after synapsis, and that in the two subsequent divisions the separation of the chromosomes can take place in four ways and with equal frequency. Consequently the germ-cells will be of eight kinds according to their chromosome constitution.

Exercise.—Make a diagram to illustrate the distribution of the chromosomes in the formation of the sex-cells when the somatic and primary germ-cells contain three pairs of chromosomes.

The primary germ-cells which are early set apart from the soma-plasm in the developing zygote rapidly increase in number and size. In the maturation stage two successive divisions of the spermatocytes and oocytes occur and four cells result, which become the sex-cells, the sperms or eggs. In the case of the spermatocytes the divisions are equal and four sperm-cells are formed, but in the oocytes the divisions are unequal with regard to cytoplasm and but one perfect egg is formed, the other three being thrown off as *polar bodies.* (See Figs. 6 and 24).

Although the nuclear divisions at maturation are indirect or *mitotic*, they differ from the usual form of mitosis.

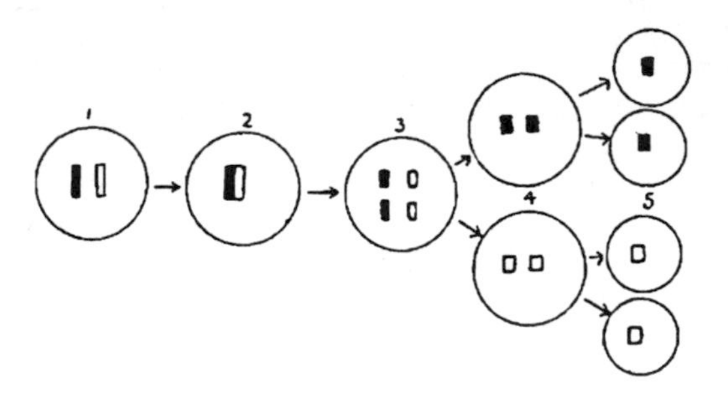

Fig. 26.—Diagram to show the phases of Reduction and the distribution of the chromosomes in the gametes. 1, spermatogonium with two chromosomes (one paternal and one maternal); 2, synapsis; 3, formation of tetrad; 4, spermatocytes; 5, spermatozoa.

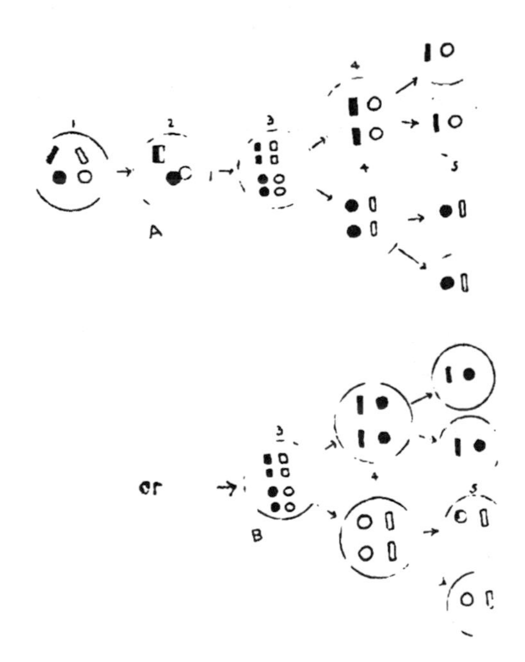

Fig. 27.—Diagram to show the phases of Reduction and the distribution of chromosomes in the gametes. 1, spermatogonium with 4 chromosomes (two paternal and two maternal); 2, synapsis; 3, formation of two tetrads; ,4 spermatocytes; 5, spermatozoa.

The essential feature is the reduction phase whereby the number of chromosomes is reduced by one-half. The modified mitotic division, called *synapsis*, consists in the union of the chromosomes in pairs (paternal and maternal) instead of splitting lengthwise, and then one member passing into a new daughter-cell, and the other member passing into the other daughter-cell.

(d)—The Chromosome Theory of Heredity

Sutton in 1904 put forward the following theory of the mechanism of heredity, based on the behavior of chromosomes.

All the cells of an individual have two sets of chromosomes obtained from the sperm and ovum respectively at fertilization. These are known as the maternal and paternal chromosomes, and they retain their separate individuality in every cell of the zygote. In some zygotes the maternal chromosomes may dominate, thus giving different physical characteristics to the developing individual.

From the following diagram (Fig. 28), where, for the sake of simplicity, the sperm and ovum have each two chromosomes, AA and BB respectively, "it can be seen that there are three possible combinations in such a domination. The two paternal chromosomes may dominate, and in that case the new individual will resemble the male parent. The two maternal chromosomes may dominate, and in that case the individual will resemble the female parent. There are two chances out of the four that the dominating chromosomes will be paternal and maternal, and in these cases the new individual will blend the characters of the two parents. It will be observed that this is exactly the ratio of Mendel's law" (Coulter).

With a form having three pairs of chromosomes there would be eight possible combinations, four pairs sixteen combinations, etc. In general ,the number of possible combinations would be represented by 2n, where *n* is the number of pairs of chromosomes in the somatic cells.

Exercise.—Try to account for the fact that no two persons are exactly alike, not even the members of one family.

In the crossing of a tall pea with a dwarf pea the chromosome explanation of the results is as follows: The former has all the chromosomes bearing the factor *tall* (T), the latter the chromosomes with the factor *dwarf* (D). The cells of the hybrid F_1, will carry a pair of chromosomes, T and D. In the maturation of the gametes of F_1 the chromosomes are reduced to one-half, and at reduction the factors will get into separate cells, thus giving two kinds of gametes. As each sex has two kinds of gametes, the possible number of combinations of zygotes will be four, viz., TT, TD, DT, DD, which agrees with Mendelian segregation.

Thus it will be seen that both the chromosomes and the factors undergo segregation.

A later extension of the theory assumes that all the factors carried by the same chromosome remain together in segregation. In other words, linkage of factors in segregation should be expected. (See page 109).

In the crossing of yellow—tall peas with green-dwarf peas, the chromosome theory explains the results as follows: The yellow-tall plants have sex-cells with the factors for tall (T) and yellow (Y) in separate chromosomes, and the green-dwarf plants factors for green (G) and dwarf (D) in separate chromosomes, but the yellow and green chromosomes are homologous; similarly the tall and dwarf chromosoes are homologous.

The cells of the hybrid of F_1 will have the chromosome constitution YGTD, the factors or determiners Y and G lying in identical positions in the one pair of homologous chromosomes, and the factors T and D in identical positions in the other pair of homologous chromosomes. All the other pairs of chromosomes carry the same set of factors.

In the maturation of the gametes of F_1, when the number of chromosomes is reduced by one-half, two types of division occur, viz., germ-cells containing the Y T and G D chromosomes, or YD and GT chromosomes. Hence there will be four kinds of gametes,—YT, YD, GT, GD.

For if the chromosomes are represented by YT and GD, the paternal being YT and the maternal by GD, then in *synapsis* only Y and G can pair together, likewise T an D. The first pair, however, operates independently of the second, so that at reduction either member of the first pair may get into a cell with either member of the second pair. Consequently the following divisions might occur:

$$\frac{YT \quad YD}{GD \quad GT}$$

This would give four kinds of gametes, each getting one or the other member of the two pairs of chromosomes, and as each sex would have these four kinds of gametes the possible number of combinations in the production of new zygotes would be 16. (Refer to page 116).

(e)—Structure of Chromosomes

Each chromosome is believed to be a linear aggregate of chromatin elements, called *chromomeres*, and is reformed at every cell division from the same chromomeres. It is probable that the chromomeres in the pairs of parental chromosomes meet in corresponding pairs and separate when the chromosomes separate. Moreover, the chromomeres in a chromosome are different, each having a definite place or *locus*. It is believed also that the chromomeres represent Mendelian factors or genes.

The recent work of Morgan and his associates with *Drosophila* reveals quite definitely the structure of the chromosome. The heritable factors, represented by loci, have a definite linear arrangement in the chromosome, and when a change occurs in some locus a corresponding change occurs in the development of the individual. Moreover, the loci act together as a system ,and a somatic change is the result of a change in the set of factors of the system. (See page 123).

The number of somatic unit characters is, therefore, a multiple of the number of chromosomes, and may be grouped according to the number of the latter.

Besides, the chromosomes have individuality, that is, they vary in size, length, and shape.

Much difference of opinion exists as to the material substance of the chromomeres. Some physiologists hold that they consist of enzymes, some hold that they are hormones, and others that they are composed of a chemical system of complex organic materials. But the opinions are as yet mere guesses.

(f)—Changes in Chromosomes

As has been already noted, chromosomes may sometimes change their constitution by 'crossing-over' of the chromomeres. They may also be modified or injured by subjecting them to certain chemicals, to high temperature, to electric currents, etc.

Mutations may arise through "crossing-over of recessive factors constantly present in the heterozygous stock." Some investigators believe that many of DeVries' Oenothera mutations arose by such a process.

Moreover, genes may not be immutable. As they are complex bodies they are liable to undergo changes "due either to the loss or addition of certain constituent atoms of molecules, or to the rearrangement of some of these."

The work of Morgan and his co-workers on eye-color in *Drosophila* would seem to bear out this suggestion. "The gene for the normal red-eye color may change so as to give rise to *blood, cherry, eosin, buff, tinged* or *white* eyes. Genes or allelomorphs that mutate in various directions give rise to what are known as *multiple allelomorphs;* hypothetically these may be explained as due to different changes, probably of a chemical nature, in a particular gene" (Conklin).

The Individual Persistence of Chromosomes.—The Chromosome Theory of Heredity takes for granted that in all the successive nuclear divisions the individual chromosomes maintain their identity.

Several investigators, including Boveri, Sutton and Morgan, have discussed evidence that the chromosomes that appear at the beginning of nuclear division are identical with the daughter-chromosomes of the preceding division, and that each of the chromosomes plays its part in the making of the organism. Frequently each may be recognized by its shape and size. Boveri's experiments with the fertilized egg-cells of a sea -urchin where he removed the nucleus of the egg-cell and a complete embryo was produced, showed that the set of chromosomes of one parent alone were sufficient for the development of a normal individual. This same investigator, however, found that if one chromosome was wanting from either the paternal or maternal set normal development would not take place.

Although the chromosomes usually disappear after division during the resting stage, in a number of cases they have been traced through the resting stage until their emergence at the next division. Wenrich has shown in *Phrynotettix* that the separate chromosomes by absorbing fluid from the cytoplasm form vesicles and unite to form the daughter nucleus. By the disappearance of the partition walls a single nucleus vesicle is formed, but under favorable conditions each chromosomal vesicle can be seen to persist through the resting stage and at the next mitosis the chromatic granules in each vesicle form a chromosome like the one from which that vesicle arose.

Summary

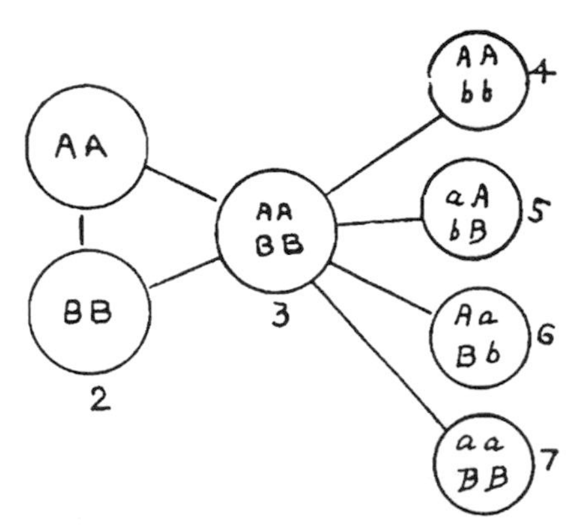

Fig. 28.—Diagram illustrating possible results of sexual fusion: 1, sperm with two chromosomes (AA); 2, egg with two chromosomes (BB); 3, fertilized egg containing two paternal chromosomes; 4, somatic cell, paternal chromosomes dominating; 5 and 6, somatic cells, one paternal and one maternal chromosome dominating; 7, somatic cell, maternal chromosomes dominating. (After Coulter).

That the germ nuclei are the bearers of the heredity qualities seems to be borne out by .

(1) the phenomenon occuring in cell-division;

(2) the phenomenon occuring in maturation;

(3) the phenomenon of fertilization;

(4) the results of breeding experiments; and

(5) the experiments of Boveri with nucleated and non-nucleated sea-urchin eggs. He found that non-nucleated framents of unfertilized sea-urchin ova could be fertilized, but such fragments developed into dwarf but normal larvæ. Again, Boveri fertilized the nucleated fragments of sea-urchin ova with sperms from another species, and obtained in a few cases dwarf larvae which showed paternal characters only. He concluded therefore that the nucleus was the exclusive bearer of hereditary qualities.

Some contradictions, however, remain to be explained:—for example:

(1) Loeb's experiments with unfertilized sea-urchin's eggs. Loeb was able to produce normal larvæ from unfertilized sea-urchin eggs by immersing them in a mixture composed of 50% 10-8n, $MgCl_2$ and 50% of sea water.

(2) Delage's experiments with non-nucleated ova of sea-urchins. Delage developed normal larvæ from non-nucleated, sea-urchin eggs by fertilizing them with sperms.

(g)—Heredity of Sex

Much progress has been made in unravelling the elementary mechanism of sex distribution both by cytological studies of the chromosomes and by studies of Mendelian inheritance.

The X or Sex Chromosome.—In the study of sexual cells of certain insects by Henking, McClung, Stevens, Wilson and others it was observed that all the eggs and half of the sperms contained an X or accessory chromosome. It was observed also that when an egg was fertilized by a sperm containing the X chromosome, a female was produced, but when an egg was fertilized by a sperm without the X chromosome, a male was produced. The female contained, therefore, two X chromosomes and the male only one. However, other observers have noted in some of the sperm-cells of certain insects, many mammals and certain plants the presence of the X chromosome, and in others a differently shaped chromosome which has been called the Y chromosome. The Y chromosome, however, does not carry any of the factors located in the X chromosome, and "does not appear to have any effect upon the development of the body characters, so that the male depends upon a single X chromosome for the development of those characters determined by the factors borne by this chromosome. The Y chromosome may, therefore be regarded as a neutral mate for the X chromosome of the male" (Babcock and Clausen). Morgan has called this the *XY type of inheritance*[1] (See also pages 111-112).

On the other hand, Doncaster and Raynor and others have observed that in *Abraxas* (English Currant Moth), chickens, ducks and canaries, *all the sperms* and *half the eggs* contain an accessory chromosome (called W to distinguish it from the accessory X chromosome of the XY type) and that the fertilized egg or zygote containing the 2 W chromosomes is a male, and the zygote with the 1W chromosome is a female. In some cases again, there appear to be evidences of a chromosome (called the Z chromosome) in some females, which may also be regarded as a neutral mate for the W chromosome of the female. Morgan calls this the *WZ type of inheritance.* (See also pages 115-116).

The two types of sex inheritance, discussed above and in chapter 15, are therefore, (1) one where the female is homogametic with two X chromosomes, or, in Mendelian terms homozygous for the sex factor, and the male heterogametic and heterozygous; and (2) one where the male is homogametic and homozygous, and the female heterogametic and heterozygous.

(1)—This type of inheritance occurs in all mammals and in a number of insects.

Equality of Sexes.

Experiments on sex-linked inheritance seem to indicate that the X chromosomes are the vehicles for the distribution of the sex factors. In both types where fertilization occurs the chances are that approximately an equal number of males and females are produced, as shown by the following diagrams:—

First type.

O = egg cell with X chromosome
S = sperm cell with an X chromosome
s = sperm cell without the X chromosome
A = Zygote with 2 X chromosomes (female)
a = Zygote with XY chromosomes (male).

		S	s	
				= male gametes
Female Gametes	O	A	a	= 2A and 2a, an equal number of males and females.
	O	A	a	

Second Type.

O = egg cell with a W chromosome
o = egg cell with a Z chromosome
S = sperm cell with a W chromosome
A = Zygote with 2 W chromosomes (male)
a = Zygote with WZ chromosomes (female)

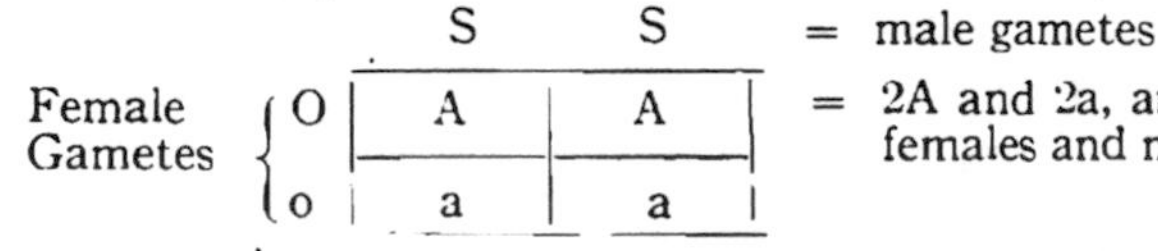

This conclusion agrees with statistical evidence which shows an approximately equal number of males and females of human birth. This and other facts already presented in chapter 15 lend support to the assumption that sex is a character dependent upon determiners or factors that act in a Mendelian manner. (See page 111).

Discuss influence of lethal factors.

It should be borne in mind that the chromosome theory of sex-determination has not been accepted by many geneticists and physiologists. The sex-chromosome has not yet been found in plants, although certain facts recently ascertained are suggestive of its presence. Recent investigations by Hertwig (1912) on frogs' eggs, by Miss King (1912) on toads' eggs, and by Riddle (1917) on pigeons showed that sex can be changed, even controlled, by varying and regulating the physiological conditions. Many physiologists claim that the "determiners" may not be confined to the chromosomes, and that certain enzymes or the known chemical specificity of the proteins and carbohydrates of the cytoplasm may possibly function as carriers or determiners. Moreover, no chemical differences can yet be detected between chromosomes carrying markedly different factors.

Parthenogenesis.—Some animals and plants are able to produce successive generations without fertilization of the egg cell. This phenomenon is called *parthenogenesis*, and is a common method of reproduction among plant-lice and gall-wasps. It also occurs in the dandelion and the hawkweed.

Parthenogenesis may be looked upon as a reversion by the sexual producing animals and plants to an asexual power possessed by their ancestors.

In parthenogenetic individuals the chromosomes of the egg are not reduced in maturation, consequently the offspring has the same number of chromosomes as the parent. It is well known that in the case of bees, wasps and ants the unfertilized eggs of the queen produce drones or males, and the fertilized eggs, females. "In all known cases of parthenogenesis the female is in the duplex condition and the male is in the simplex, or partially duplex, condition. The female in all cases has the greater chromatin content." (Castle).

Several experimenters have been able to cause the unfertilized egg to develop artificially by some physical change or from the action of chemical substances. (page 141).

(h)—The Control of Sex

For ages breeders have looked for a method of controlling sex. While recent studies seem to point to the Mendelian character of sex, the predetermination of sex still eludes the investigator. Many theories have been advanced but most have been discredited on rigid examination.

The more important theories may be grouped as follows:

1. Those involving purely internal and organic factors:
   (a) The predominance of one parent over the other with reference to *vigor* and *age*. But opinions differ here as to the expected sex of the offspring. Some cattle breeders maintain that the young bulls beget a majority of males and the old bulls a majority of females. Some breeders believe that sex

is dependent on the *age of the ovum* at conception; in other words, when conception occurs early in heat most of the calves are heifers; late in heat bulls. Other breeders are of the opinion that sex is dependent on the *age of the spermatozoa*, fresh spermatozoa producing females.

(b) The changes occurring during *maturation* of the egg. (See page 133).

(c) The influence of the *nervous system*.

2. Those involving purely external factors:

(a) The influence of *nutrition*, as in the case of plant-lice, queen bees, and tadpoles.

(b) The influence of *temperature, moisture, seasons*.

3. Those involving the existence of two special groups of *chromosomes* following Mendel's Law of Heredity (page 81).

This is the view most commonly held by geneticists.

4. Those involving the influence of various factors, and it is probable that several of the factors mentioned above may be concerned. It is possible, for example, that "environmental changes may slightly disturb the regular working out of the two possible combinations that give male or female. Such disturbances may effect the sex ratio but have nothing to do with sex determination" (Castle).

Some recent experiments suggest, however, that sex-ratio can be influenced to some extent by selection. Miss Helen D. King's investigations with albino rats brought out the interesting fact that by selecting (1) a pair from litters that showed the highest proportion of females, and (2) a pair from litters showing the highest proportion of males the sex ratio can be changed. Selection was caried on in the two strains for about 25 generations. As a result the proportion of females to males in the inbred female-producing strain was 100 to 82,0. and that in the in-bred male-producing strain was 100 to 117. (See Journ, Exp. Zool. 1918, 27).

While no breeding experiments have been carried on with cattle with the direct object of producing female-producing strains, many cases occur where bulls get more female than male calves.

The study of twins offers some hints in the matter of determination of sex. Ordinary human twins are often unlike both as to sex and to physical traits like ordinary brother and sister. Each originated from a separate ovum and has its own fetal membrane. It is evident, therefore, that external factors, since these were alike, had no influence in determining the sex.

"Identical twins" are always alike in regard to sex and physical traits. It is believed that they originated from one ovum for they are enclosed in the same fetal membrane. At a very early cleavage the parts become separated and each develops a complete individual. The sex was presumably determined in the fertilized ovum before separation.

The Free Martin.

Cattle breeders have observed for a long time that when twins of different sexes are produced the female, the free-martin, is usually sterile. Prof. F. Lillie attributes the phenomenon to the action of the hormones on the two fetuses which have a constant interchange of blood on account of the development of wide arterial and venous anastomosis. If the twins are both males or both females the hormones produce no harm, but if one is male and the other female the reproductive system of the female is largely suppressed.

Chapter 17 — IN-BREEDING

(Consult *In-breeding and Out-breeding* by East and Jones)

Considerable difference of opinion exists among breeders as to the effect of In-breeding, that is of the mating of closely related members of a family.

Line-breeding restricts the mating of individuals to a single line of descent. Among plants many self-fertilize, such as wheat, oats, etc., and maintain a vigorous existence. Darwin's experiments with pansy, morning glory, cabbage lettuce, buckwheat and beets (see pages 77-78) went to show, however, that self-fertilization tends to weaken the off-spring and crossing to add vigor and fertility.

It has been observed, moreover, that "close in-breed- is always detrimental to racing ability. Few winners have been produced by matings of animals more closely related than half-cousins" (Watson, *Heredity*).

Among higher animals, however, it is impossible to bring about as close fertilization as that occurring in self fertilizing plants, as none are hermaphrodite.

Mendelism throws light on the question. It shows that in-breeding in itself is not necessarily injurious, but great care must be exercised to prevent injury. It is clear that as organisms become more closely related they tend to become homozygous, and many become practically so through long in-breeding. Their characters exist in the duplex or double-dose condition, as in Pure Lines. If the essential characters in both parents are strong and without defects the result of in-breeding will be the strengthening or fixing of these same characters. The duplex dose of determiners develops prepotency in all characters, good and bad alike. The most important characters relate to vigor and fertility, and the breeder must see to it that these are in no way impaired by in-breeding.

While undoubted evidences of prepotency, or the superior power of some animals of impressing characters on their offspring, are quite abundant, a full analysis of the factor concerned has not yet been made. It is probable, however, that the analysis will be along Mendelian lines, as suggested above, and that prepotency results from (1) *in-breeding* where homozygous dominants tend to be produced; (2) *assortative mating*—where like mates with like in regard to certain characters; (3) the chance presence of a *large number of dominant traits* in one individual; (4) sex-linkage and crossing over; (5) the action of *multiple allelomorphs;* or (6) variations and interrelations of the factors concerned.

If, however, some of the essential characters are weak in the parents the progeny of in-breeding will show this weakness intensified.

In-and-in-breeding has developed the greatest breeds of horses, cattle, swine and sheep. It is a fact that the race-horse is more nervous and more delicate of constitution than the draft-horse, but in-breeding did not produce these apparently undesirable qualities. The quality of nervousness has been selected as essential to a good race horse.

In nature there are vigorous self-fertilizing and cross-fertilizing flowering plants, but the latter preponderate.

The reason for this preponderance, according to East and Shull, is that the *vigor* of "a plant is influenced not only by the characters it possesses but also according to whether these characteristics are present in the pure or in the hybrid condition. If, therefore, self-fertilized strains or varieties have survived competition in the struggle for existence they are inherently stronger than cross-fertilized strains or varieties" (East, Bul. 243. U.S. Bureau of Plant Industry).

Miss H. D. King inbred, brother and sister, two series of albino rats for 25 generations. The productiveness did not decrease, nor did the constitutional vigor. The space of life in both sexes was increased (Journ. Exp. Zool; 1918, 26).

Exercise.—Show by Mendel's formula that when a heterozygous pair is self-fertilized, the progeny becomes after several generations more and more homozygous. Make a graph showing the reduction of heterozygous individuals in successive generations.

The following chart illustrates line-breeding in a noted Clydesdale strain;

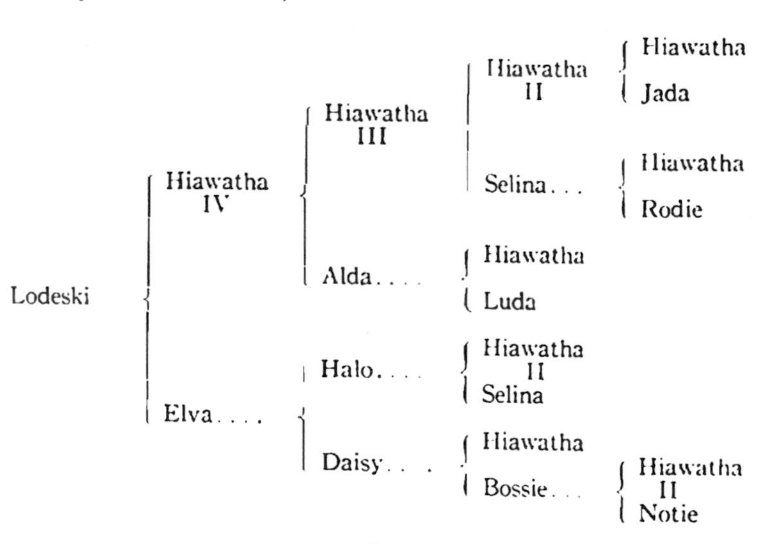

(from *Journal of Heredity*)

Lodeski is line bred to Hiawatha. She is not bred in line to Jada but rather to her grandsire Hiawatha whose blood has been perpetuated and intensified.

The following chart illustrates line-breeding in a noted Shorthorn strain;

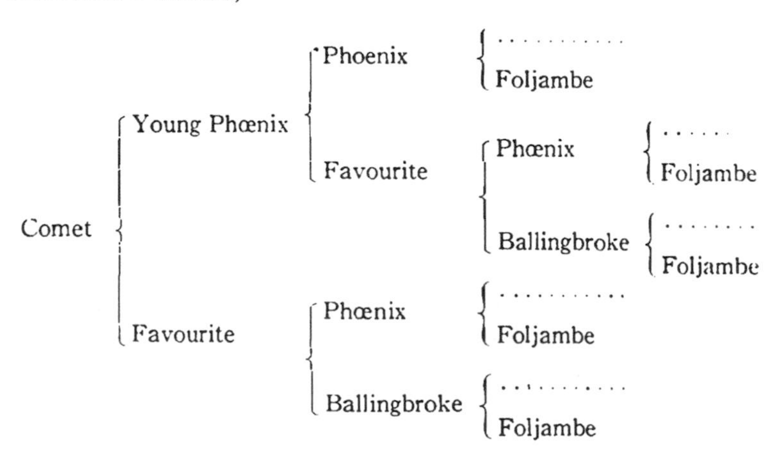

Account for the extinction of the Bates-Duchess line of Shorthorns.

Observations on in-breeding with both plants and animale frequently reveal the fact that fertility is reduced in the first two or three inbred generations. On the other hand, Castle states that he inbred *Drosophila*, the pomace fly, for fifty-nine generations without any apparent effect on either the vigor or the fecundity. Nevertheless when he crossed two inbred strains the offspring were superior in productiveness to either inbred strain (Compare Shull & East's *Experiments with Corn*, page 78).

Bos inbred rats for six years; for the first three years no ill effects were observed, but after that a rapid decline in vigor and fertility occurred.

Castle again states that "the heterozygous yellow mouse is a vigorous lively animal, but the homozygous yellow mouse is so feeble that it perishes as soon as produced, never attaining maturity. Cross-breeding has, then, the same advantage over close-breeding that fertilization has over parthenogenesis. It brings together differentiated gametes, which, reacting on each other, produce greater metabolic activity" (Castle, *Heredity*.)

In relation to man most countries have laws preventing the marriage of close relatives, but in others, notably Egypt and Abyssinia, the custom permitted of close inter-marriages. Some recent studies go to show that in certain communities where close inter-marriages were practised for two or three generations the mental, moral and physical condition of the people was below normal. This result was likely due to the lack of care in the choice of mates and to the fact that defectives were allowed to marry defectives.[1]

Co-efficient of In-breeding.—Galton's law of Ancestral Heredity has been utilized by some students of pedigrees to ascertain the extent of inbreeding, i.e. the factor of the the number of times that the same animal appears in the pedigree. Pearl has coined the term "Co-efficient of Heredity" as a precise measure of inbreeding (Bull. 215 Maine Agric. Exp. St. 1913). This co-efficient expresses the relation between the maximum and the actual number of different ancestors, the formula being

$$Z_n = \frac{100\,(P_{n+1} - Q_{n+1})}{P_{n+1}}$$

In the pedigree of Lodeski (page 148) the formula would work out as follows:

$$Zo = \frac{100\,(2 - 2)}{2} = 0 \text{ per cent.}$$

$$Z_1 = \frac{100\,(4-4)}{4} = 0 \text{ per cent.}$$

$$Z_2 = \frac{100\,(8-5)}{8} = 37\tfrac{1}{2} \text{ per cent.}$$

In the pedigree of Comet (page 149).

$$Zo = 0 \text{ per cent.}$$

$$Z_1 = \frac{100\,(4 - 3)}{4} = 25 \text{ per cent.}$$

$$Z_2 = \frac{100\,(8 - 4)}{8} = 50 \text{ per cent.}$$

(1)—Galton was of the opinion that the Athenians of the period 530-430 B.C. were the most gifted in history, when from 45,000 free-born males surviving the age of fifty came 14 of the greatest men of all time. According to Galton also only 250 men per million of the population of Great Britain became eminent. The Athenians practised endogamy, and marriage with aliens was punishable by law. To this selective inbreeding some would attribute the development of the high gifts of the Athenians of the Age of Pericles. On the other hand, there is much evidence in support of the theory that racial mixing or outbreeding within limits makes for greater ability than inbreeding in the case of isolated countries.

Castle doubts the utility of such a co-efficient, inasmuch as vigor and fecundity have no necessary relation to the amount of in-breeding in the ancestry, and as uniformity is not necessarily increased when a heterozygote is used as an ancestor. Moreover, Galton's Law is not now considered a safe guide in biological reasoning (Castle, *Genetics and Eugenics*).

Co-efficient of Relationship.—It is quite possible, however, for two animals with the same *Co-efficient of Inbreeding* to differ greatly in germinal constitution, for example when a closely inbred animal of one breed is mated to another closely inbred animal of another breed. To give some measure of the inter-relation of the lines of descent, Pearl has devised the term *Co-efficient of Relationship*, which is the per cent of the individuals in each line which are also represented in the other line, and is a measure of the community of ancestry of the dam and the sire (Pearl, *Inbreeding and Relationship Co-efficients. Am. Nat. 1914*).

King Melio Rioter 14th, the Jersey Bull, has a co-efficient of Inbreeding of 90% at the seventh ancestral generation, but a co-efficient of Relationship of 40%; while Blossom's Glorena had co-efficients of 93% and 0% respectively.

"These two co-efficients taken together, give us the first quantitative measure of inbreeding as a system of mating, but obviously they do not tell anything concerning the actual germinal constitution of any individual resulting from a given system of inbreeding. The germinal composition of any individual can be determined only by actually testing its breeding qualities, its transmissive powers. But an indication of the germinal constitution of an individual produced by any long-continued system of inbreeding, as far as the degree of heterozygosity or homozygosity is concerned, can be obtained by the laws of probability to Mendelian formulæ" (East and Jones).

We have already seen (page 96) that the probable character of the segregating generation when inbred, when n character pairs are concerned, may be expressed by the expansion of the binomial $(3:1)^n$. For any generation it may be expressed by the expansion of the binomial $[(2^r-1)^n+1]$, where the exponent of the first term gives homozygous characters and that of the second term the number of heterozygous characters.

As East and Jones point out, "this reduction applies only to the whole population, or to a representative sample of the population, in which every member is selfed, in which each individual is equally fertile, and in which all the progeny are grown in every generation. In practice in an inbreeding experiment, usually only one individual in self-fertilization or two individuals in brother and sister mating are used to produce the next genertion.

Thus the rate at which complete homozygosity is approached depends on the constitution of the individuals chosen" (*Inbreeding and Outbreeding*).

Chapter 18—A COMPARISON OF PLANT AND ANIMAL BREEDING

The plant breeder has an advantage over the animal breeder since he can handle thousands of individuals to one by the latter. In selection he is, therefore, better able to find variant forms, even mutations, that suit his purpose. Moreover, in hybridizing better opportunities are offered for the segregation of the unit characters.

The factors of rapidity of production, the possibility of asexual reproduction, and the relative cheapness of individuals are also important advantages in favor of the plant breeder.

On the other hand, the fact that plants are influenced to a greater extent by soil and climate than animals are, and that it is more difficult to keep actual records in the case of plants, places the plant breeder at a disadvantage.

The plant breeder endeavors to produce new breeds or races, different in some important characters from their type. On the other hand, the animal breeder does not strive after the production of new breeds or races, but rather "improvements within the breed or race, such as increased size, greater milk production, improved beef quality, increased fecundity, or some such quality."

(a)—Animal Improvement

The various *pure-breeds* among horses, cattle, sheep, swine, etc., have arisen from native stock by selection and prevention of mixing with inferior strains. (See pages 33-34).

In recent years the introduction of *Pedigree Registers* has done much to keep the breeds pure. Such Registers arose as a necessity on account of the danger of introduct-ion of impure blood, where impure individuals were common, and the great loss attending such introduction.

In late years *Advanced Registers* have been formed in which *Performance* is the basis of registration.

Unrecorded animals may be (1) *scrubs*, those with mixed strains, and (2) *unregistered*, or those of pure strains but whose records have been lost, and whose ancestry in us-known.

Systems of Breeding.—In general four systems are practised:

(1) *Mixed breeding*, where no attention is paid to ancestry. In this system evidently no improvement can take place or be expected.

(2) *Pure breeding*, where only registered animals are used.

(3) *Grading*, where only the sire is pure-bred. By care it is possible to build up a sound herd without much trouble or expense. With every succeeding generation the animals are becoming more and more pure-bred. For economic production such high-grade animals are equal to pure-breed.

Discuss Co-operative Breeding.

(4) *Crossing*, where pure breds of different races are crossed. This system is not recommended except for the modification of a breed ,with the chance of getting an improved form.

"Breed *in* to fix type; breed *out* to secure vigor; in genreal, compromise "sums up fairly accurately the experience of breeders.

It will be observed that the splendid achievements of animal breeders have been made without a knowledge of genetics as it is understood by the scientists of to day. The lore of the breeder's craft is to a large extent the valuable body of experience accumulated through long centuries after repeated failures and successes. It differs from genetics in that it gets results through the adoption of traditional rules or methods while the latter seeks for the natural laws that underlie and explain the methods. In this way genetics often breaks down old erroneous ideas, suggests new methods, and makes for future progress in the art of breeding. The best results, of course, will be secured when the breeder is thoroughly acquainted with both the art and the science of breeding.

(b)—Plant Improvement

In general the three main lines that have been adopted for the improvement of plants are:

1. Continued *selection* of superior individual plants (*Line Selection*).
2. *Separation* from mixtures of such individuals as show desirable qualities and breed true to type (*mutations*).
3. Combination of the desirable qualities of two strains or varieties by *hybridization*.

There are two methods of growing plants for selection, which Webber calls the Nursery Method and the Field Method. By the former "each plant is grown under the most favorable conditions for its best development." By the Field Method "the selections are made from plants grown under normal field conditions.' In order to keep the selected strain up to a high standard, selection must be continued year after year, for unless this is done the plants gradually revert to the normal or average type of the strain before selection began.

The second line of improvement, viz., the separation of mutations showing desirable qualities, although unconsciously used to some extent for centuries, is only now being adopted by breeders, as a result of the studies of DeVries, Bateson, Nilsson, and Burbank. Darwin recognized mutations as sports. A practical difficulty exists in our inability to distinguish them from variations of the ordinary sort. The only test is to breed them. Mutations come true to type and do not show any tendency to revert. Ordinary variations are of value mainly in the production of improved strains of a race, which are soon lost when the selection is discontinued. Mutations or sports, on the contrary, are of value in the production of distinctly new races and varieties which maintain their new characters without continued selection."

Methods of Improvement.—Seven distinct methods of improving old types of cultivated plants and establishing new ones may be recognized:

1. *Hallett's Method.*—Plants are grown under the best possible conditions, and the best are selected. This methof was first used by Hallett who originated several superior strains of wheat, oats and barley (See Chapter 5). It pro-

ceeds on the assumption that the characters acquired by some of the plants on account of favorable environment are inherited. The origin of the new forms is to be explained probably on the assumption that the favorable conditions allowed certain germinal characters to find expression.

2. *Rimpau's Method.*—Plants are grown in a mass under ordinary or unfavorable conditions and the best selected. Rimpau believed that such a procedure revealed best the heritable characters in the true genotype. When such selection is practised year after year pure lines tend to be produced which, we have seen, are stable.

3. *DeVries' Method.*—This method consists in the selection of desirable mutation forms that may arise in plantings.

4. *Vilmorin's Method.*—The improved sugar beet has arisen by this method. The seeds of selected plants are sown in separate plots, and at maturity one or more samples of each plot is tested for sugar content. Seed is secured from the plants of the plot that has given the best sample.

5. *Johannsen's Method.*—Consists in the isolation of desirable Pure Lines by planting individual seeds from superior mother plants until stable pure lines are obtained.

6. *Burbank's Method.*—where hybridization and selection are practised on a large scale, with the expectation that new forms may arise through the "random result of fortuitous combinations."

7. *Mendel's Method.*—where hybridization is practised with the expectation of getting the desired combination by definite crosses. (See Mendelism).

Registered Seed and Seed Centres in Canada.—A system of selection and registration of seed exists in Canada under the supervision of the Canadian Seed Growers' Association. The grades leading up to "Registered Seed" are:—

1. *Hand Selected Seed,*
2. *Improved Seed,* and
3. *Elite Stock Seed.*

Hand Selected Seed is cleaned seed obtained from heads, etc., which are uniform in character and which have been selected by hand from sound, vigorous and normally developed plants.

Improved Seed is seed originating from hand-selected seed, but which is not yet entitled to public recognition as "Registered Seed."

Elite Stock Seed is (1) the product of hand selected seed plots after at least three years' selection; or (2) pure stock originating from a single plant, the progeny proving satisfactory in plot or field tests.

"Registered Seed" is the progeny of Elite Stock Seed, handled in accordance with the rules of the Association. Most of the Registered seed is now produced by *Seed Centers*, usually groups of farmers who multiply Elite stock seed according to the rules of the Association.

The third line of plant improvement, viz., by *Hybridization*, has been in use for over 150 years, but no general principle had been established until Mendel published a report in 1865. Until 1900 no person seemed to be able to predict with any degree of certainty the result of crossing varieties of plants (See Mendelism, pages 81-87).

Selection of Fluctuations

Sugar Beets.

The sugar content of sugar beets has been gradually raised from 4-6 per cent. to 16-18 per cent. during the nineteenth century. The credit for the improvement is mainly due to the methods of analysis and selection introduced by Louis Vilmorin. For the last thirty-five years, however, no perceptible inprovement has been made in spite of rigorous selection of mother beets, and it would seem that the strain is kept at the maximum range of fluctuation only by the selection and isolation of the best plants.

High Protein and High Oil Corn.

The Illinois Agricultural Experiment Station during the years 1896-1915 increased the protein content of corn seed from 10.92% to 14.53% and the oil content from 4.70% to 8.46% by selection. (See Babcock and Clausen for an admirable discussion of the Illinois results).

Methods of Breeding Timothy, Wheat, Potatoes, etc.

It is impossible here to describe in detail the method of breeding improved strains of grasses and cereals.

Reference may be made to Cornell University bulletins 251 and 313 by Dr. Webber, and to L. H. Newman's "Plant Breeding in Scandinavia."

Canadian Grains

Canadian Experiment Stations have originated several grains of superior merit. The Marquis is a productive, early ripening variety of spring wheat, relatively free from rust and yielding a fine quality of flour. It is especially recommended for the Western prairie provinces. It originated as a cross between the *Hard Red Calcutta* (female) and the *Red Fife* (male), made in 1892 by Dr. A. P. Saunders, and separated by Dr. Chas. Saunders in 1904.

The *Huron* wheat, a valuable variety for the Eastern provinces, originated from a cross between the *Ladoga* and *White Fife*. The *Prelude*, a very early variety and adapted to northern districts, is derived from four varieties by crossing—Gehun, Ladoga, White Fife and Hard Red Calcutta. Its origin may be represented as follows:

Ladoga x White Fife
|
Alpha x Hard Red Calcutta
|
Fraser x Gehun
|
Prelude

The *Early Red Fife* was derived from the Red Fife.

The *Manchurian Barley*, a selected strain of Mensury, is a heavy yielder.

Dr. Chas. Saunders[1] originated a new and valuable hulless oat which he has called the *Liberty Oat*.

The *Arthur* Pea originated from a cross between the Mummy and the Multiplier varieties.

The preceding varieties were produced at the Dominion Experimental Farms.

Mr. Seager Wheeler of Rosthern, Sask., originated in 1910 the *Red Bobs* from Bobs, an Australian wheat, and in 1911 the *Kitchener* from Marquis, both strains showing great promise. (See Buller, *Essays on Wheat*).

The Ontario Agricultural College has also originated several new varieties. Among the more important are the *O.A.C. No. 21* Barley, a selected strain of Mandscheuri six-rowed bearded, relatively free from rust and very

(1)—Dr. Chas E. Saunders was born in London, Ontario, in 1867; graduated from Toronto (1888) and Johns Hopkins Universities (1891); Assistant at the C. E. F.; appointed Dominion Cerealist in 1903; has published many reports, bulletins and papers on Cereals and Cereal Breeding.

productive the *O.A.C.* No. 72 *Oats*, a selected strain of the Siberian oat, white with a pinkish tinge, of fine quality and a high yielder; the *O.A.C. No. 3 Oat*, a selected strain of the Daubeney, and an early variety.

Macdonald College has developed by selection the Quebec Yellow Corn, an early strain of flint corn, the No. 12, 11 strain of Yellow Intermediate mangel, and the No. 90, 11 strain of the Purple Topped Bangholm swede.

Canada has also produced several valuable varieties of fruits. The McIntosh Red Apple originated as a seedling of the Fameuse, and the Herbert raspberry as a seedling of the Cuthbert.

Grain Breeding in Sweden

The "single ear" method of Le Couteur and Shireff, the British breeders, has been adopted by the Swedish breeders at Svalof as the basis of their selection for hybridizing experiments. They had, however, the advantage over the early breeders of a knowledge of Mendel's work. According to the law of Segregation, they knew that the second generation (F_2) contains all the types, resulting from a cross. Their procedure was as follows: "Two known sorts are crossed and the whole progeny from all second and succeeding generations is sown together *en masse*" (Newman).

By this method the breeder and Nature together eliminate the unfit combinations and select the superior types. For several generations, five or more, the cultures are "rogued", when preliminary tests for yield are made and the field trials begun.

At Svalof, therefore, new varieties of wheat have been originated by selection and crossing. "Pansar" and "Fylgia" are the best varieties in use at the present time. The following diagram shows their origin.

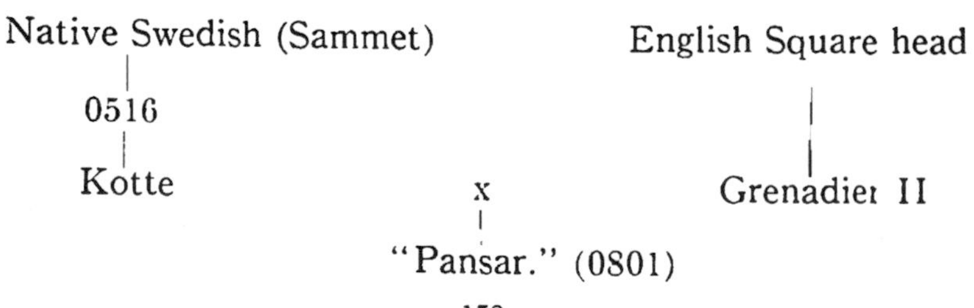

English Sq. hd. — Leutewitzer Sq. hd. — English Sq. hd.

Extra Sq. hd. I x Grenadier II

Tystofte Smaahvede (Danish) x Extra Square head II

"Fylgia." (0860)

The following table shows the improvement made in the wheat crop at Svalof in a comparatively short time, the values being given as percentages of the yield of Native Swedish wheat:—

Pansar (Kotte x Grenadier II)	140
Fylgia (Smaahvede x Extra squarehead II)	135
Tystofte Smaahvede	131
Sol	131
Extra-squarehead II (Extra squarehead I x Grenadier II)	129
Kotte	128
Grenadier	121
Tystofte Standhvede	121
Criewemer 104	121
Vilhelmina	115
Extra-squarehead I	112
Native Swedish (Sammet)	100
English Squarehead	99

Four or more pairs of factors are undoubtedly concerned in the yield averages of the Swedish wheats. Nilsson-Ehle mentions the following: resistance and non-resistance to rust (R, r). to winter (W,w), to lodging (L, l),and tillering and non-tillering (T, t). It is evident that those wheats that combine the greatest number of good factors such as R.W.L.T. will yield highest, and those that have the most poor factors such as r.w.l.t. will yield lowest. The plan adopted was to select and cross varieties with the highest yield and to select for a number of years the best for propagation. Thus higher and higher yielding varieties were obtained as higher and higher yielding parents were crossed.

Breeding Varieties of Plants that are Disease-Resistant

With the great advances in methods of plant-breeding in recent years the problem of breeding plants resistant to disease has appealed to many workers The investigations of Biffen of England showed that it was possible to breed by crossing a strain of wheat resistant to Yellow Rust (*Puccinia glumarum*), but all attempts to breed a strain resistant to Black Stem Rust (*P. graminis*) have met with disappointment. Recent researches go to show that this latter rust exists in America as six or more distinct physiological races or biologic forms and strains differing in their virulence, and that the nature of rust-resistance is not yet thoroughly understood. A variety of wheat resistant in one region to rust is sometimes susceptible in ano.her region.

The inheritance of rust-resistance depends, therefore, upon the specific relation between host and fungus. It is likely, therefore, that by selection, rather than breeding, resistant strains of wheat will be found for limited areas, which are capable of resisting the local biological strains of rust. (Read E. C. Stakman's Article on Resistance to and Biologic Forms of Black Stem Rust. in Journ. Agric. Research, 1915 and 1917).

In general, the resistance of plants to other fungous diseases is specific, depending upon the strain of variety and upon the physiological differences between strains of the fungus. Orton has developed a watermelon (*Citrullus vulgaris*) that is resistant to the wilt disease (*Fusarium niveum*), and a strain of cotton resistant to the cotton wilt (*F vasinfectum*). Hansen has produced hybrid pears that are largely resistant to pear blight (*Bacillus amylovorus*).

Prof. R. H. Biffen's experiments on the breeding of a strain of wheat that was at the same time hard, high-yielding and resistant to rust (*Puccinia glumarum*) are interesting and suggestive.

First he determined that *hardness* was dominant *to softness* by crossing Red Fife with Rough Chaff, an English wheat. In the F_2 generation he was able to obtain a form that combined hardness with the high-yielding quality.

Next, Biffen crossed Red Rivet an English wheat, very resistant to rust, with Michigan Bronze, a strain of American wheat very susceptible to rust. The F_1 generation showed that susceptibility was dominant to immunity, but in the F_2 generation one fourth of the plants were found to be immune.

Similar results were obtained when he crossed American and Michigan Bronze.

Biffen also crossed *Squarehead's Master* (rust susceptible) with a Ghurka wheat (rust resistant) and obtained a strain resistant to P. glumarum, called *Little Joss.* This strain is now grown very extensively in England. (Read Biffen's *Systematized Plant Breeding* in Science and the Nation, 1917).

Chapter 19—EUGENICS AND EUTHENICS

In the section entitled "Heredity and Inheritance," attention was called to Nature and Nurture as factors that influence the organism. By Nature was meant the Inheritance, and by Nurture the result of the environmental influences, such as climate, food supply, functional activities, etc.

Natural Selection undoubtedly played an important part in man's early period of development, when there was a constant struggle for existence with wild animals, his surroundings, and his own fellow men. In these modern days, however, man has almost freed himself from the sway of Natural Selection, as we ordinarily understand it. Many agencies now exist specially for the care of persons who would have perished in the struggle of existence in days of savagery. The weak are protected, and there is no special premium on strong, healthy individuals. There is, it is true, a process of social selection in operation, but this does not serve the purpose of Natural Selection.

Much study has been given in recent years to devising plans for the improvement of society. These may be grouped into two types:

1. the improvement of the human breed; giving rise to the science of *Eugenics;* and
2. the making of beautiful and wholesome surroundings and the selection of heathful and educative occupations; giving rise to the science of *Euthenics.*

In Great Britain and other countries the population, we are told, is being recruited from their inferior stock, for 50% of the children born are produced by one-eight of the total adult population. Pearson has investigated the number per family of many grades and kinds of stock in Great Britain and has found that the highest (7.0) occurs in pathological stock, and the lowest in well-to-do and professional classes (1.5–2.0). As at least four children to a family are required, under average conditions, to maintain its numbers undiminished from generation to generation, it can be easily seen that the inferior stock is being maintained and increased.

There is practically no hindrance to the rapid multiplication of the unfit and degenerate. Recent studies show that good and bad physique, the liability to and the immunity from disease, the moral characters, and the mental temperament are inherited in man, and with much the same intensity."

It is inevitable that unless more stringent regulations regarding marriage are adopted our nation stands exposed to "the gravest risks of retrogression."

By long-continued wars old Rome lost her best men and the less fit were left to propagate the race. "Homo" replaced "Vir," and the nation fell a prey to the more virile northern conquerors. The wars of Louis XIV and Napoleon weakened France for generations, since the young able-bodied men were drafted into the army and perished, and the relatively more unfit left at home.

Eugenics, in a word, attempts to apply the same principles to man as the breeder applies in the development of his herds of fine stock. The subject is one that deserves the careful thought of every citizen who has at heart the welfare of his country.

Eugenics was a term introduced by Francis Galton[1] in 1883 in his "Inquiries into Human Faculties" to express "the study of the agencies under social control that may improve or impair the racial qualities of future generations either physically or mentally." Galton in the same work also stated the aim of Eugenics: "Its first object is to check the birth rate of the unfit instead of allowing them to come into being, though doomed in large numbers to perish prematurely. The second object is the improvement of the race by furthering the productivity of the fit, by early marriages and the healthful rearing of their children".

The Eugenic idea is an old one. Plato discussed the problem in the *Republic* and the *Laws*, in which he advanced many of the suggestions being advanced nowadays for the improvement of the human race, such as regulation of marriages, taxation of bachelors over 35 years of age, and care of mothers and children.

About the beginning of the 17th century Campanella discussed the problems of Eugenics, much after the fashion

(1)—Sir Francis Galton was a cousin of Chas. Darwin. He took his medical degree at Cambridge. He travelled extensively in Africa, and published two books on his experiences. He studied anthropology and heredity, and endowed a research fellowship in the University of London for the Study of Eugenics. His publications are—"Hereditary Genius" (1869), "Human Faculty and its Development" (1883), "Natural Inheritance" (1889).

of Plato, in *The City of the Sun.* To Francis Galton, however, we are indebted fór bringing Eugenics to the attention of the public as a plan worthy of serious consideration and, whenever practicable, of immediate application.

The Eugenist believes that while much can be done to improve the human race by euthenic methods,—by education, control of dieases, sanitary improvements, and general conditions of living—there remains much to be done through the application of improved methods of breeding. It has already been shown that Nature is probably a more important factor than Nurture in the highest development of organisms, and that the improvement of the common races of animals amd plants has followed along well recognized lines of the laws of inheritance.

It is hardly to be expected that the method of the stock breeder for race improvement, viz., the selection of the best for breeding purposes will be applied in the case of man. The development of a race of *supermen* is utopian, and the most that can be hoped for is the elimination of the worst human kinds from the possibility of reproduction.

The question of the relative infertility of the more well-to-do and socially efficient stock has given rise to much serious discussion. Many causes have been assigned for the low birth-rate which leads to *race-suicide.* In some cases the condition is undoubtedly due to the lack of healthy outdoor work and the absence of the "simple life," in other cases it is due to "selfish celibacy and selfish non-maternity," and in others to the high cost of living which makes it hard to rear a family under modern conditions. (See Kellicott p. 115).

Mendelism of Human Characters

Recent investigations go to show that the laws of inheritance apply also to man, and that many characters, —good and bad—are transmitted according to Mendelian fashion. Eugenics, however, has to deal specially with the problem of preventing the transmission of undesirable traits such as feeble-mindedness, insanity, epilepsy, deaf-muteness, etc., all being heritable. (See Goddard, *Feeble-mindedness; Its Causes and Consequences*, 1914).

Not much direct evidence has been accumulated regarding the inheritance of normal characteristics, but it has been shown that eye-color (blue and brown), hair color, hair texture, skin color, nervous temperament, etc., are inherited according to Mendel's laws.

With regard to the study of inheritance of abnormalities and disease, however, considerable progress has been made, and, strange to say, some of these are dominant to the normal. (See page 168).

Moreover, a study of pedigrees of certain families discloses the fact that undoubtedly bad and undesirable moral and mental characters, as well as good and desirable

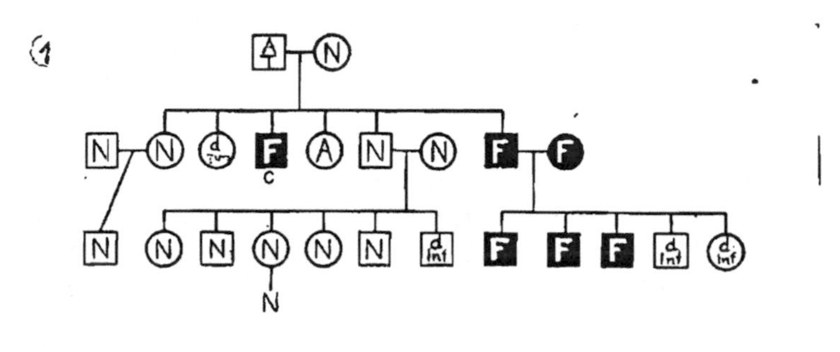

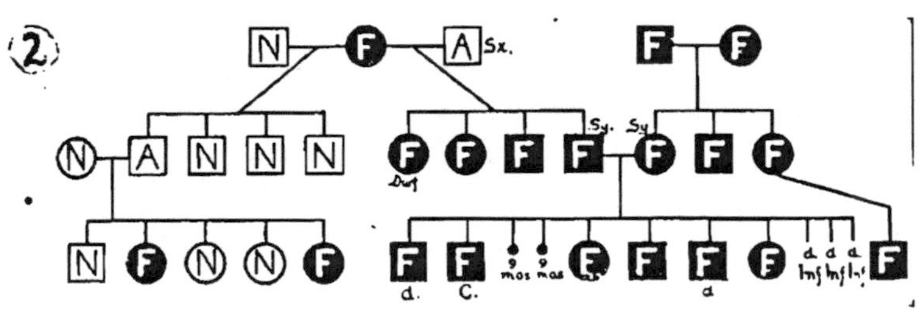

Fig. 29.—In these diagrams the circle represents a female, the square a male. N inside a circle means normal; F on a black background means feeble-minded; A, alcoholic; T, tubercular; Sx, sexually immoral; Sy, having syphilis.

ones, often akin to genius, are inherited. (Read the story of the "Jukes," the "Kallikak," the "Nam," the "Edwards," the "Darwin," the "Bach," the "Zero," the "Hill Folk " the "Pineys" and other families).

A critical study of many family records has been made in recent years by Dr. H. H. Goddard, Dr. C. B. Davenport[1] and others in the United States, and by the Galton Laboratory for Research in England. The results are such as to convince the thoughtful of the terrible consequences

(1)—Prof. C. B. Davenport, the noted American experimental biologist and geneticist, was born in 1866 and educated at the Brooklyn Polytechnic and at Harvard. He acted as instructor in Zoology at Harvard (1888-1889), and was Professor of Zoology, Chicago University (1899-1904), and is now Director of the Biological Laboratory, Cold Spring Harbor, L. I. He has published several valuable works in Experimental Zoology.

following the marriage of feeble-minded persons. Feeble-mindedness, moveover, may possibly reveal itself in several forms: addiction to alcohol; criminalistic, immoral, and epileptic tendencies. The diagrams (Fig. 29) show two family records and are worthy of careful study. In the first a normal woman married an alcoholic tubercular man. In their family of six two were feeble-minded (F), one was alcoholic (A), one died young, and two were normal (N). One of the feeble-minded sons married a feeble-minded woman, and of the five children three were feeble-minded, and two died in infancy. (Fig. 29, 1).

In the second diagram a feeble-minded woman married twice, first to a normal man, and second to an immoral man.

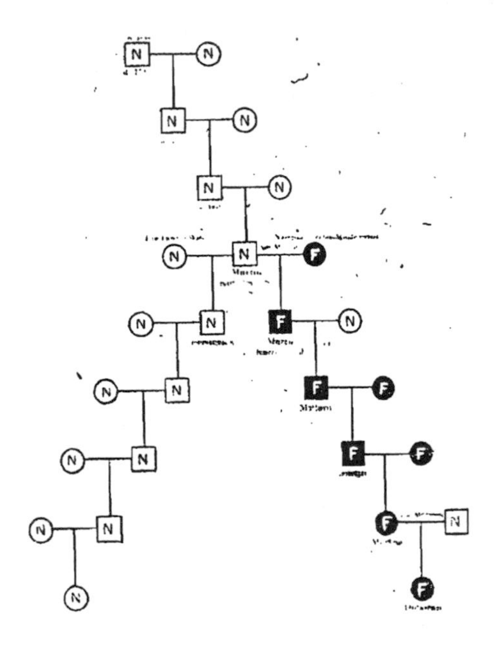

Fig. 30.—Diagram outlining history of Kallikak family. Squares stand for males, circles for females; N for normal people, F for feeble-minded. Five generations on the side of the feeble-minded girl contain 480 individuals—143 feeble-minded, 33 immoral, 24 drunkards, and 3 epileptics. On the normal side are 496 descendants, none of whom are feeble-minded. (After Goddard.)

The record of the family of the second marriage was a terrible one, all being feeble-minded. A similar terrible record was noted in the next generation. (Fig. 29, 2).

We give herewith the inheritance in six families, three bad and three good.

The **Kallikak Family** (Goddard). Fig. 30)

Line A.

In five generations 480 direct descendants from a normal father and feeble-minded mother have been accounted for as follows:

143 known to be feeble-minded.
291 mental status unknown or doubtful.
36 illegitimate.
33 sexually immoral, mostly prostitutes.
24 confirmed alcoholics.
3 epileptics.
82 died in infancy.
3 criminals.
8 keepers of disreputable houses.
46 known to be normal.

Line B.

In five generations 496 descendants from the same normal father as in Line A and a normal mother have the following record:

All but one of normal mentality.
Two men known to be alcoholic.

One case of religious mania.
Among the rest have been found nothing but good representative citizenship, numbering doctors, lawyers, educators, judges, traders, etc.
No epileptics or criminals.
Only fifteen children died in infancy (Guyer).

The **Jukes Family** of Sullivan Co., N.Y. (Dugdale)

Jukes born 1720: had five daughters.

In five generations about 1200 persons including idle, ignorant, lewd, vicious, pauper, diseased, imbecile, insane, and criminal specimens. 300 died in infancy; 310 paupers in alms-houses; 440 wrecks—diseased; 130 criminals; 60 thieves; 7 murderers. Cost to society over 2½ millions of dollars.

The **Zero Family.** A Swiss family (Jorger).

A terrible example of inheritance of degeneracy.

The founder of this family was "the result of two generations of intermarriage, the second tainted with insanity. He married an Italian vagrant wife of vicious character. Their son married a member of a German vagabond family, and this pair had seven children, all characterized by vagabondage, thievery, drunkenness, mental and physical defect, and immorality (See Guyer, *Being Well Born*).

The **Edwards Family** in United States (Winship).

Of 1394 descendants of Jonathan Edwards and Elizabeth Tuttle,—in 1900—295 were college graduates; 13 college presidents; 65 professors; 60 physicians; 100 clergymen, etc; 75 Army and Navy officers; 60 authors and writers; 100 lawyers; 30 judges; 1 V.P. U.S.; 3 U.S. Senators.

The **Darwin Family** in England (Whetham). (See chart).

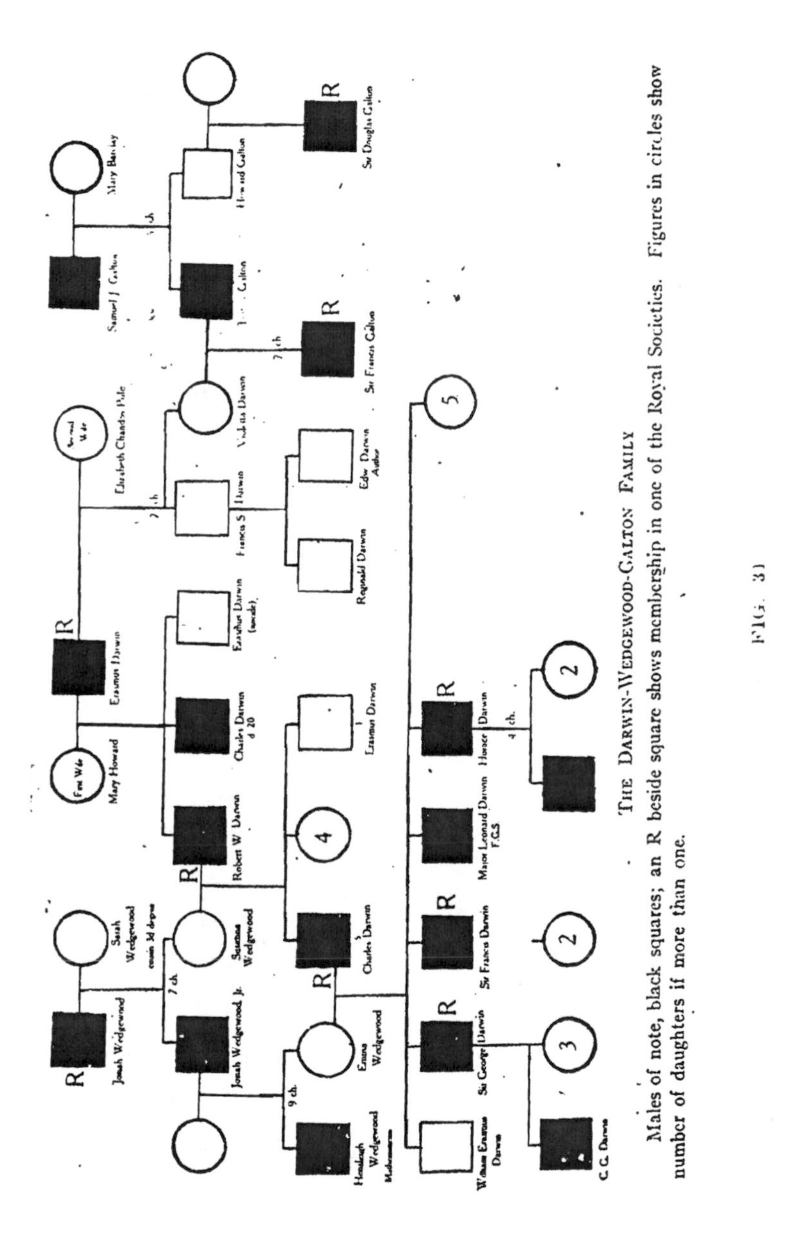

THE DARWIN-WEDGEWOOD-GALTON FAMILY

Males of note, black squares; an R beside square shows membership in one of the Royal Societies. Figures in circles show number of daughters if more than one.

FIG. 3)

A good example of the inheritance of scientific ability.

The **Bach Family** of Thuringia.

In six generations appeared 57 musicians of note.

(See chart p. 12, Downing).

In the study of many human traits "it is often very difficult to say whether they are dominant or recessive, whether they are controlled by one or by several factors, whether they are sex-linked or independent" (East and Jones).

Castle proposes the following grouping of human characters: (1) clearly Mendelian; (2) Mendelian and sex-linked; (3) probably Mendelian but dominance imperfect or uncertain; (4) blending; and (5) hereditary but to what extent or how, uncertain.

Examples of (1) and (2) are the following:—

Dominant	Recessive.
Brown or black eye	Grey or blue eye
Black hair	Red hair
Curly hair	Straight hair
Brachydactyly	Normal
Polydactyly	Normal
Congenital cataract	Normal
Night blindness	Normal
Color blindness	Normal—Sex-linked
Deaf-mutism	Normal—Sex-linked
Hypotrichosis	Normal
Diabetes insipidus	Normal
Glaucoma	Normal
Huntingdon's Chorea	Normal
Normal	Albinism
Normal	Feeble-mindedness[1]
Normal	Hæmophilia (bleeding)—Sex-linked
Normal	Hereditary hysteria
Normal	True dwarfism
Normal	Multiple sclerosis
Normal	Suxceptibility to cancer
Normal	Susceptibility to tuberculosis

As examples of (3) are hare-lip, left-handedness, tendency to produce twins; of (4), skin color in negro-white crosses, shape of head in mixed races; of (5) general mental, musical, literary, artistic and mechanical ability, cretinism, congenital deafness, some forms of epilepsy and insanity, longevity, etc.

(1)—East estimates that the factor for feeble-mindedness occurs in one person out of fourteen of the population of the U. S.

Alcohol and Heredity

Reliable data regarding the effect of alcohol on offspring are hard to procure, and often furnish contradictory conclusions. Certain facts are known, however, which may be suggestive rather than conclusive:

1. Nicloux has shown that the reproductive glands of certain mammals have a strong affinity for alcohol, and may contain almost as large a proportion as the blood itself.

2. The reproductive glands of inebriates often show atrophy and other degenerative changes, but sometimes more children are born to alcoholics than to normal parents.

3. Offspring of inebriates often show serious abnormalities, such as "rickets, dwarfism, predisposition to tuberculosis and epilepsy, and to crime and mental diseases": (Forel). Stockard's experiments with guinea-pigs showed clearly the injurious effects of alcohol on offspring. On the other hand, Pearl's experiments on the inherited effects of alcohol on chickens showed fewer but stronger progeny.

Many investigators are of the opinion, however, that feeble-mindedness of children is not the result of the alcoholism of the parents, but that the alcoholism of the parents is usually a symptom of feeble-mindedness or some form of lack of control.

Experiments with Guinea Pigs to Test Influence of Alcohol on Germ Cells

	Number of matings	Abortive	Died at Birth	Lived
Alcoholic				
Alcoholic males and normal females	24	19	7	5 (all runts)
Normal males and alcoholic females	4	2	2	2
Alcoholic males and alcoholic females	14	13	1	0
Control				
Normal males and normal females	9	0	0	17 (all vigorous)

Influence of Alcohol on Progeny

Dogs (Hodge)

	Alcoholic Pair	Normal Pair
Number of Whelps	(7-7-6-3)23	(5-3-8-8-5-6-3-7)45
Deformed	(2-3-3-0)8	(1-0-0-2-0-0-0-1)4
Born Dead	(2-2-2-3)9	(0-0-0-0-0-0-0-0)0
Viable	(4-0-0-0)4	(4-3-8-6-5-6-3-6)41
	(17.4%)	(90.2%)

Men (Demme)

	Ten Alcoholic Families	Ten Normal Families
No. of Children	57	61
Deformed	10	2
Idiotic	6	0
Epileptic, choreic	6	0(2bkw.)
Died at Birth	3	3
Normal, viable	10 (17%)	54 (88.5%)

Venereal Diseases and Heredity:- The alarming prevalence of syphilis and gonorrhoea calls for action in the prevention of marriages of tainted persons. Although these diseases are not heritable in the true sense of the term, yet their occurrence amounts almost to a certainty that the children will contract them. Infantile blindness and pelvic disorders of women follow gonorrhoea, while paresis, locomotor ataxia, and arterio-sclerosis, and feeble-mindedness of the children follow syphilis. It would seem advisable that medical inspection before marriage should be enforced for the sake of the children and the welfare of the state.

In order to prevent the transmission of undesirable germ plasm the following proposals have been made by eugenists:—

a. More stringent marriage laws,
b. Sexual segregation of defectives,
c. Stricter control of immigration, and
d. Measures of sterilization of dangerous defectives by vasectomy and oophorotomy.

From the positive euthenic standpoint much inprovement might be effected by reforms in social conditions that prevent marriage of desirables, by a campaign to enlighten the public regarding the ideals of eugenics, by the encouragement of marriage between individuals above the average physically, mentally, and morally, by the reduction of infant mortality, and by financial aid to parents toward the support of their children.

"The idea of selection for parenthood as determining the nature, fate and worth of living races is Darwin's chief contribution to thought, and finds in Eugenics its supreme application" (Saleeby).

1. Discuss the selective action of war, social regulations and customs regarding marriage, and disease "which may improve or impair the inborn qualities of future generations either physically or mentally."

2. Discuss the probable causes of the declining birth-rate in most countries.

3. Discuss the statement, 'Education is nothing more than the giving or withholding of opportunity.'

4. Discuss the question of inter-racial marriages from an eugenic standpoint.

5. Discuss "Human Conservation".

6. Is school instruction in sex problems and sex hygiene advisable? (Read Bigelow's *Sex Education*).

7. "The lessons which the Eugenist seeks to enforce are written in flame across every page of zoology: the wiping out of less perfectly developed and less adaptive tribes by better equipped ones is going on daily under our very eyes" (McBride). Discuss.

Chapter 20 — IMPORTANT EVENTS IN THE HISTORY OF GENETICS

333 B.C. Aristotle observes the chick in the shell, and states definite views regarding embryology in "De Generatione Animalium."

1660. Robert Hook discovers the cellular structure of plants.

1672. Malpighi describes the development of the chick.

1677. Leeuwenhoek discovers spermatozoa.

1694. Camerarius discovers sex in plants.

1719. Fairchild crosses plants artificially—"Fairchild's Sweet William."

1751. Linnæus publishes the *Systema Naturæ*.

1759 Wolff in his *Theoria Generationis* discards the *preformation theory* of Malpighi and Bonnet and proposes the *developmental theory* of the embryo.

1760. Kolreuter produces the first plant hybrid in a scientific experiment.

1786. Spallanzani shows that sperms are essential to fertilization.

1793. Sprengel discovers insect fertilization of plants and dichogamy.

1809. Lamarck in *La Philosophie Zoologique* expounds the theory of *Use and Disuse* and the *Inheritance of Acquired* Characters.

1817. Pander discovers the three primary germ-layers of the embryo.

18 7-8 Von Baer discovers the mammalian ovum and lays the foundation of comparative embryology.

1831. Brown describes the nucleus in plant cells.

1835. Dujardin discovers and describes *sarcode* (protoplasm).

1838-9. Schwann and Schleiden formulate their *cell-theory*.

1841. Kolliker demonstrates the cellular origin of sperms in the testes.

" Remak proposes the theory that every cell comes from a pre-existing cell by a process of division.

1843. Martin Barry observes the union of sperm and ovum in the rabbit.

1846. Von Mohl brings the term *protoplasm* into general use.

1859. Charles Darwin expounds the *Theory of Natural Selection* in the *Origin of Species*.

1861. Max Schultze recognizes the cell as composed of protoplasm and a nucleus.

1863. Fritz Muller formulates the *Recapitulation Theory*, first stated by Von Baer.

1864. Herbert Spencer in the *Principles of Biology* suggests the existence of "physiological units' in the reproductive elements.

1865. Gregor Mendel publishes the results of experiments in crossing strains of garden peas and enunciates the laws of hybridization.

1866. Hæckel in his *Generelle Morphologie* emphasizes the material continuity of offspring and parent.

1869. Darwin in *Variation of Animals and Plants under Domestication* puts forward the hypothesis of *pangenesis*.

1875. Hertwig determines that fertilization results from the union of *one* egg and *one* sperm.

1880–1. Francis Balfour publishes his *Comparative Embryology*.

1883. Francis Galton uses the term *Eugenics* in *Inquiries into Human Faculties*.

—— Van Beneden and Boveri discover chromosomes.

—— Van Beneden shows that every nucleus may contain nuclear substance obtained from each parent, and that a reduction in the number of chromosomes occurs in maturation.

1884. Carl Nageli presents his *Mechanico-physiological Theory* of Evolution.

1889. Hugo DeVries puts forward the theory of *Pangens* or "Intracellular Pangenesis."

—— Galton in *Natural Inheritance* enunciates the *Laws of Ancestral Inheritance*.

1891. Henking finds two kinds of spermatozoa in *Pyrrhocoris*.

1893. Weismann publishes his *Germ Plasm Theory*.

1895. Eimer proposes his *Orthogenetic Theory of* Evollution.

1896. E. B. Wilson publishes *The Cell in Development and Inheritance*.

1900. Re-discovery of Mendel's publication (1865) and the publication of similar investigations by Bateson, DeVries, Correns and Tschermak.

1901. DeVries publishes "*Die Mutations Theorie*."

—— McClung finds a numerical difference in the chromosomes in the spermatozoa of the grasshopper.

1902.	Bateson and Cuenot first apply Mendelism to animals.
——	Sutton, (W. S.) puts forward The Chromosome Theory of Heredity.
1903.	Johannsen puts forward his *Pure Lines* Theory.
——	Formation of the American Breeders' Association.
1905.	Stevens and Wilson determine the correct relation of the *accessory* or *X chromosome* to sex.
1906.	Bateson and Punnett discover linkage in heredity.
1907.	The Galton Eugenics Laboratory founded in London.
1909.	Bateson publishes "Mendel's Principles of Heredity."
1914.	Goddard discovers that feeble-mindedness is a recessive unit character.
1915.	Morgan *et al* publish "The Mechanism of Mendelian Inheritance."
1910-1919	Investigations and discoveries of Nilsson, Nilsson-Ehle and Witte in Sweden; Johannsen in Denmark; DeVries in Holland; Vilmorin, Delage and others in France; Bateson, Biffen, Darbishire, Durham, the Gartons, Hu st, Lock, Pearson, Punnett, Saunders, Schuster, Sutton, Wilson and others in Great Britain; Burbank, Castle, Davenport (C. B.), East, Emerson, Jones, Loeb, Morgan, Pearl, Shull, Tower, Wilson and others in the United States; Correns, Baur, Klebs, von Lochow and others in Germany.

Chapter 21—IMPORTANT LITERATURE

Babcock and Clausen: Genetics in Relation to Agriculture, McGraw-Hill Book Company, 1918.

Bailey and Gilbert: Plant Breeding, Macmillan, 1915.

Bailey, H. L. The Survival of the Unlike, Macmillan.

Bateson W. Mendel's Principles of Heredity. Cam. University Press, 1909.

Bigelow, M. A.: Sex Education. Macmillan, 1919.

Buller, A. H. R.: Essays on Wheat, Macmillan, 1919.

Castle, W. E: Genetics and Eugenics, Harvard Univ. Press, 1912.

Coulter, Castle, etc.: Heredity and Eugenics. Univ. Chicago Press, 1918.

Coutler and Coulter: Plant Genetics, Univ. Chicago Press, 1919.

Darbishire, A. D: : Breeding and the Mendelian Discovery, Cassell, 1911.

Darwin, Chas.: Origin of Species, Appleton.

" ": Cross and Self Fertilization in the Vegetable Kingdom, Chaps. I and XII. Appleton.

" ": The Variation of Animals and Plants under Domestication. Appleton.

Davenport, C. B.: Heredity in Relation to Eugenics. H. Holt, 1911.

Davenport, E.: Principles of Breeding. Ginn & Co. 1907.

" ": Domesticated Animals and Plants. Ginn & Co. 1910.

Delage and Goldsmith: The Theories of Evolution. Huebsch, N.Y. 1912.

DeVries, Hugo: Species and Varieties. Open Court. 1906.

" ": Plant Breeding. Open Court, 1907.

Downing, E.R.: The Third and Fourth Generation. Univ. Chicago Press, 1918.

Dugdale, R. L.: The Jukes: A study on Crime, Pauperism, Disease and Heredity. Putnam's Sons. 1877.

East, E. M.: The Relation of certain Biological Principles to Plant Breeding. Bull. 158. Conn. Agric. Expt. Station.

East & Jones: Inbreeding and Outbreeding. Lippincott, 1919.

Estabrook A. H.: The Jukes in 1915. Carnegie Inst. 1916.

Goodrich, E. S.: Evolution of Living Organisms. Jack.

Goddard, H. H. : The Kallikak Family. Macmillan. 1912

" " Feeble-mindedness: its causes and consequences. 1914.

Guyer, M. F. : Being well Born. Bobbs-Merrill Co., Indianapolis, 1916.

Johannsen, W. : Elemente der Exakten Erblichkeitslehre. Fischer. Jena, 1913.

Jordan, D. S. : The Human Harvest. Boston, 1907.

Kellicott W. E. : General Embroylogy. Holt & Co. 1913.

Kellogg, Vernon L. : Darwinism To-day. Holt & Co. 1907.

Lock, R. H : Variation, Heredity and Evolution. Murray, 1906.

Locy, W. : Biology and its Makers. Holt & Co. 1st ed. 1908.

Loeb, Jacques : The Organism as a whole. Putnam's. 1916.

Mitchell, Chalmers : Evolution and the War. Murray. 1915.

Morgan, T. H. Evolution and Adaptation. Macmillan. 1903.

" " Heredity and Sex. Columbia Univ. Press, 1913.

' " Critique of the Theory of Evolution. Princeton Univ. Press, 1916.

" ' The Physical Basis of Heredity. Lippincott, 1919.

" et al The Mechanism of Mendelian Heredity. Holt & Co. 1915.

Newman, L.H. : Plant Breeding in Scandinavia. Can. Seed Growers' Association, Ottawa. 1912.

Osborne, H. F. : From the Greeks to Darwin. Macmillan.

" " The Origin and Evolution of Life. C. Scribners' Sons N.Y. 1918.

Parker, G. H. : Biology and Social Problems. Houghton Mifflin & Co. 1914.

Pearl, R. : Modes of Research in Genetics.

Punnett, R. C. : Mendelism, 4 ed. Macmillan, 1916.

Schuster, E. : Eugenics. The Nation's Library. 1912.

Seward A. S. : Darwin and Modern Science. Cam. Univ. Press. 1909.

Thomson, J. A. : Heredity. Murray, 1908,

Thomson & Geddes: Evolution. Williams and Norgate.

Walter, Herbert E. : Genetics. Macmillan, 1913.

Watson, Jas. : Heredity. T. C. & E. C. Jack, London 1912.

Webber H. J. : Plant Breeding for Farmers. Bull. 241, Cornell Agric. Expt. Sta. 1908.

Weismann, August: The Germ Plasm, Scott. London, 1893.

" " The Evolution Theory, 2 vols, Arnold, London, 1904.

Wilson, Jas. : A Manual of Mendelism. A. & C. Black, 1916.

Excellent articles may be found in the Annual Reports of the American Breeders' Association, The Journal of Heredity, The Journal of Genetics, The American Naturalist, The Biological Bulletin, Science, Genetics, etc.; The U.S. Year Book and Experiment Station Bulletins, and The Eugenics Review of London University, England.

GLOSSARY (From Various Sources)

Acquired Character.—A modification of bodily structure or habit which is impressed on the organism in the course of the individual life.

Allelomorphs.—Factors occuring in the same locus in homologous chromosomes, and for this reason producing "contrasting" or "alternative" characters.

Amphimixis.—The mingling of heredity units of two parents in sexual production.

Autosome.—Any other chromosome than the sex-chromosomes.

Biometry.—The branch of science dealing with the statistical investigation of organic differences.

Bud Mutation.—A mutation occurring in the very early history of a bud such that a branch is produced which differs genetically from the remainder of the plant.

Chimera.—A mixture of tissues of different genetic constitution in the same part of a plant.

Chromomeres.—The chromatin granules, which are sometimes arranged like the beads on a necklace.

Chromosomes.—Term applied to certain minute bodies in the nucleus of the animal and vegetable cell, which appear at definite periods in the division of the cell, and are constant in number for each species of animal or plant.

Clone.—A group of individuals produced from a single original individual by some process of asexual reproduction such as division, budding, slipping, grafting, parthenogenesis (when unaccompanied by a reduction of the chromosomes), etc.

Crossing-over.—Exchange of chromatin material between homologous chromosomes.

Cytology.—The branch of biology which treats of cells, especially of their internal structure.

Cytoplasm.—That portion of the protoplasm of the cell outside the nucleus.

Diploid.—The number of chromosomes normally found in the somatic cells of a species; twice the gametic or haploid number.

Endosperm.—The substance stored in a seed adjacent to the embryo for its early nourishment.

F_1, a symbol introduced by Mr. Bateson, as an abbreviation for the first hybrid generation.

F_2, a symbol for the second hybrid generation.

F_3, a symbol for the third hybrid generation.

Factor.—An independently inheritable element of the genotype by the presence of which some particular character in the organism is made possible; gene; determiner.

Fertilization.—The union of male and female sex cells.

Fluctuations (Fluctuating Variations.)—The slight differences normally found in organisms and attributed *either* to environmental influences *or* to recombinations of genetic factors.

Gene.—See factor.

Genotype.—The constitution of an organism with respect to the factors of which it is made up; the sum of all the genes of an organism.

Germ-plasm.—That part of the cell-protoplasm which is the material basis of heredity and is transferred from one generation to another.

Gamete.—A name for the reproductive cell, whether male or female, in both animals and plants.

Haploid.—The number of chromosomes normally found in the gametes of an individual; one-half the somatic or diploid number.

Heterozygosis.—The condition of an organism due to the fact that it is a heterozygote.

Heterozygote.—A zygote resulting from the union of two gametes bearing *dissimilar* factors—one a dominant, the other a recessive.

Homozygote.—A zygote resulting from the union of two gametes bearing *similar* factors, which may be either both dominant, producing a dominant homozygote or both recessive, producing a recessive homozygote.

Hormone.—A substance secreted or found in some organ or tissue and carried thence in the blood to another. organ or tissue which it stimulates to functional activity or whose functions it inhibits.

Hybrid.—The offspring of animals or plants of different genotypes, varieties, species or genera.

Lethal.—Destructive of Life.

Linkage.—That type of inheritance in which the factors tend to remain together in the general process of segregation; "gametic coupling" of the older terminology.

Locus (*pl. loci*).—A definite point or region in a chromosome at which is located a genetic factor or gene.

Multiple Allelomorphs.—Factors occupying the same locus of homologous chromosomes.

Mutant.—An individual of a genotypic character differing from that of its parent, or those of its parents, and not derived from them by a normal process of segregation or by crossing-over.

Ontogeny.—The development of the individual as opposed to phylogeny.

Phenotype.—The sum of the externally obvious characters of an individual or a group of individuals.

Phylogeny.—The history of the evolution of a species or group, distinguished from ontogeny.

Pure Line.—A group of individuals derived solely by one or more self-fertilizations from a common homozygous ancestor.

Prepotency.—The property said to be possessed by certain individuals, especially amongst stallions and bulls, of transmitting their qualities to their offspring whatever female they are mated with.

Reduction Division.—One of the last two divisions in gametogenesis, when homologous chromosomes are dissociated and pass into different gametes.

Recessive Characters.—Those which, in a cross between individuals the two characters of each of which bear one of the same Mendelian pair, entirely disappear in the first hybrid generation.

Reversion.—The production, on crossing, of one or more characters of a supposed remote ancestor of the two forms crossed.

Segregation.—The reappearance in definite ratios, in the second hybrid generation, of the characters of two forms crossed; and of the first hybrid generation (where this differs from the dominant character).

Sex-chromosome.—The accessory chromosome which has come to be associated with one or the other sex, or one member of a pair of morphologically or physiologically distinct chromosomes which carry a factor or factors for sex.

Sex-linked.—Applied to factors located in the sex-chromosomes or to the characters conditioned by them.

Sex-ratio.—The proportion of males and females in a population.

Soma.—The body, as opposed to the germ cells.

Synapsis.—The conjugation of maternal and paternal chromosomes preceding the reduction of division.

Unit Characters.—Characters which behave as units in heredity.

Zygote.—The result of the union of two gametes (See Gamete).

Index

H

I

J

K

L

M

CPSIA information can be obtained at www.ICGtesting.com
Printed in the USA
LVOW13s1146071013

355765LV00002B/271/P